Undergraduate Texts in Mathematics

Editors

S. Axler
F.W. Gehring
K.A. Ribet

Springer
New York
Berlin
Heidelberg
Barcelona
Hong Kong
London
Milan
Paris
Singapore
Tokyo

Robert J. Valenza

Linear Algebra

An Introduction to
Abstract Mathematics

 Springer

Robert J. Valenza
Department of Mathematics
Claremont McKenna College
Claremont, CA 91711
USA

Mathematics Subject Classification (1991): 15-01, 13-01, 20-01, 47-01

Library of Congress Cataloging-in-Publication Data
Valenza, Robert J., 1951–
    Linear algebra: introduction to abstract mathematics / Robert J.
Valenza.—[1st e.]
        p.   cm. — (Undergraduate texts in mathematics)
    Includes bibliographical references and index.
    ISBN 0-387-94099-5 (alk. paper)
    1. Algebras, Linear.   I. Title.   II. Series.
QA184.V35   1993   93-26143
512′.5—dc20

Printed on acid-free paper.

Production managed by Karen Phillips, manufacturing supervised by Vincent Scelta.
Camera-ready copy prepared from the author's Microsoft Word for Windows files.
Printed and bound by R.R. Donnelley and Sons, Harrisonburg, VA.
Printed in the United States of America.

9 8 7 6 5 4 3 (Corrected third printing, 1999)

ISBN 0-387-94099-5 Springer-Verlag New York Berlin Heidelberg
ISBN 0-387-94099-5 Springer-Verlag Berlin Heidelberg Berlin   SPIN 10706145

To my father,
who through his own example
taught me how to work,
and to my mother
in loving memory

Preface

These lecture notes, which took shape from a one-semester course taught to sophomores and juniors at Claremont McKenna College, constitute a substantial, abstract introduction to linear algebra. Although we use material from elementary calculus on occasion to illustrate major ideas, there are virtually no formal prerequisites. This is not to say that the material is easy. Many students who have never needed to make much effort in previous mathematics courses will find themselves seriously challenged.

What is the nature of linear algebra? One might give two antipodal and complementary replies; like wave and particle physics, both illuminate the truth:

THE STRUCTURAL REPLY. Linear algebra is the study of vector spaces and linear transformations. A vector space is a structure which abstracts and generalizes certain familiar notions of both geometry and algebra. A linear transformation is a function between vector spaces that preserves elements of this structure. In some sense, this discipline allows us to import some long-familiar and well understood ideas of geometry into settings which are not geometric in any obvious way.

THE COMPUTATIONAL REPLY. Linear algebra is the study of linear systems and, in particular, of certain techniques of matrix algebra that arise in connection with such systems. The aims of the discipline are largely computational, and the computations are indeed complex.

This text leans heavily, even dogmatically, toward the structural reply. In my experience in both pure and applied mathematics, the recognition that a given problem space is a vector space is often in itself of more value than any associated computation. Moreover, realistic computational problems are almost exclusively done by computer, and therefore incessant hand-drilling in matrix techniques is both redundant and maladroit. Finally, linear algebra as abstract, noncomputational mathematics is often one's first encounter with mathematics as understood and appreciated by mathematicians. I hope that the student will learn here that calculation is neither the principal mode, nor the principal goal, nor the principal joy of mathematics.

Throughout, we emphasize structure and concept over calculation. The lecturer will note that the first three chapters bear the names of fundamental categories. Here are some further examples:

(i) Early on, we explicitly introduce and use the language and most basic results of group theory.

(ii) A few diagram chases are suggested, one of which yields a particularly elegant proof of the change of basis formula.

(iii) The proof of the associative law for matrix multiplication is postponed until after the representation of linear transformations by matrices is introduced. At this point it becomes simply an aspect of the associativity of composition of functions.

(iv) The equality of the column and row ranks of a matrix is demonstrated entirely through the formal properties of the dual space. Accordingly, we define the transpose not just for matrices, but for arbitrary linear transformations, and we fully reconcile these two notions.

An outline of the exposition follows.

(1) SETS AND FUNCTIONS. We begin by reviewing notation and terminology, most of which should be familiar, in one form or another, from early calculus. The equivalence of bijectivity and invertibility of functions is the main result of the first half of the chapter. A brief digression on cardinality follows; this is not used at all in the sequel, but does provide a brief and appealing respite from the stream of formalities. The chapter concludes with an introduction to the symmetric group on n letters. This material is, of course, used later in the discussion of determinants and leads gracefully into the next, more radical topic.

(2) GROUPS AND GROUP HOMOMORPHISMS. Abstract groups are admittedly nonstandard fare for a course in linear algebra, but the rationale for their inclusion here is, perhaps ironically, as much pedagogical as mathematical. On the mathematical side, a vector space is first an additive group, and a linear transformation is first a homomorphism of additive groups. Moreover, group theory plays a critical role in matrix theory. Thus we lay the foundations here for most of what follows, and what follows is made simpler thereby.[1] On the pedagogical side, one must recognize that the comprehension and composition of proofs is the central means by which one encompasses abstract concepts. The group, in all of its axiomatic austerity, provides a wonderful training ground for dealing with informal axiomatic systems. The student will at first lack intui-

---

[1] We do not venture beyond that which is mathematically prerequisite to the remainder of the text; for example, the basic counting theorems for finite groups are not included.

tion, but in learning to write correct proofs, this is not altogether a disadvantage. There are three sections. The first defines groups and subgroups and develops a few fundamental properties. The second introduces group homomorphisms, and already many linear algebraic themes begin to appear. (For instance, the characterization of the inverse image of an element under a homomorphism.) The last section briefly discusses rings and fields. The terminology is used throughout (rings sparingly), but the reader who takes *field* to mean either the real or complex numbers, will suffer no serious consequences.

(3) VECTOR SPACES AND LINEAR TRANSFORMATIONS. Building on additive group theory, this chapter introduces the two central objects of linear algebra, with some key examples. (The lecturer will no doubt want to supplement the material with many pictures.) We develop the basic arithmetic properties of vector spaces and subspaces, the notions of span and spanning sets, and the fundamental properties of linear transformations. Throughout, the proofs remain straightforward, and against the background of the previous chapter most students find little difficulty here.

(4) DIMENSION. The topics covered include linear dependence and its characterizations, basis and its *many* characterizations, and the fundamental structure theory of vector spaces: that every vector space admits a basis (*Vector spaces are free!*) and that every basis of a given space has the same cardinality. The student will see explicitly that the vector space axioms capture two primary concepts of geometry: coordinate systems and dimension. The chapter concludes with the Rank-Nullity Theorem, a powerful computational tool in the analysis of linear systems.

(5) MATRICES. Within this chapter, we move from matrices as arrays of numbers with a bizarre multiplication law to matrices as representations of linear systems and examples *par excellence* of linear transformations. We demonstrate but do not dwell on Gauss-Jordan Elimination and *LU* Decomposition as primary solution techniques for linear systems.

(6) REPRESENTATION OF LINEAR TRANSFORMATIONS. This long and difficult chapter, which establishes a full and explicit correspondence between matrices and linear transformations of finite-dimensional vector spaces, is the heart of the text. In some sense it justifies the structural reply to those who would compute, and the computational reply to those who would build theories. We first analyze the algebra of linear transformations on familiar spaces such as $\mathbf{R}^n$ and then pass to arbitrary finite-dimensional vector spaces. Here we present the momentous idea of the matrix of a transformation relative to a pair of bases and the isomorphism of algebras that this engenders. For the more daring, a discussion of the dual space follows, culminating in a wholly noncomputational and genuinely illuminating proof that the column and row ranks of a matrix are equal. Finally, we discuss transition matrices (first formally, then computa-

tionally), similarity of matrices, and the change of basis formula for endomorphisms of finite-dimensional spaces.

(7) INNER PRODUCT SPACES. The emphasis here is on how the additional structure of an inner product allows us to extend the notions of length and angle to an abstract vector space through the Cauchy-Schwarz Inequality. Orthogonal projection, the Gram-Schmidt process, and orthogonal complementation are all treated, at first for real inner product spaces, with the results later extended to the complex case. The examples and exercises make the connection with Fourier Series, although we only consider finite approximations.

(8) DETERMINANTS. The determinant is characterized by three fundamental properties from which all further properties (including uniqueness) are derived. The discussion is fairly brief, with only enough calculation to reinforce the main points. (Generally the student will have seen determinants of small matrices in calculus or physics.) The main result is the connection between the determinant and singularity of matrices.

(9) EIGENVALUES AND EIGENVECTORS. Virtually all of the threads are here woven together into a rich tapestry of surpassing texture. We begin with the basic definitions and properties of eigenvalues and eigenvectors and the determination of eigenvalues by the characteristic polynomial. We turn next to Hermitian and unitary operators and the orthogonality of their eigenspaces. Finally, we prove the Spectral Decomposition Theorem for such operators—one of the most delightful and powerful theorems in all of mathematics.

(10) TRIANGULATION AND DECOMPOSITION OF ENDOMORPHISMS. The discussion is a somewhat technical extension of the methods and results of the previous chapter and provides further insight into linear processes. (In a one-semester course, only the most energetic of lecturers and students is likely to alight on this turf.) In particular, we cover the Cayley-Hamilton Theorem, triangulation of endomorphisms, decomposition by characteristic subspaces, and reduction to the Jordan normal form.

With deep gratitude I wish to acknowledge the influence on this work of two of my teachers, both of whom are splendid mathematicians. Professor Wilfried Schmid taught me much of this material as an undergraduate. While my notes from his class have long since vanished, my impressions have not. Professor Hyman Bass, my thesis advisor at Columbia University, taught me whatever I know about writing mathematics. Many of his students have remarked that even his blackboards, without editing, are publishable! These men have a gift, unequaled in my experience, for the direct communication of mathematical aesthetics and mathematical experience. Let me also thank my own students at Claremont McKenna College for their patient and open-minded efforts to encompass and to improve these notes. No one has ever had a more dedicated group.

Two of my best students, Julie Fiedler and Roy Corona, deserve special praise. While enrolled in my course, they proved to be such astute proofreaders that I asked them both to assist me through the final draft of the manuscript. Although I had expected that they would simply continue their proofreading, they quickly progressed beyond this, offering scores of cogent suggestions serving to clarify the material from the student's perspective. I hope that they recognize and take pride in the many changes their comments have effected in the final product.

The aesthetic affinity of mathematics and music has always been powerful—at least for mathematicians. At times we compose, and at times we conduct. Linear algebra is one of the great symphonies in the literature of mathematics, on a par with the large works of Beethoven, Brahms, or Schubert. And so without further delay, the conductor raises his baton, the members of the orchestra find their notes, and the music begins...

Contents

SUPPLEMENTARY TOPICS

Index of Notation

1

Sets and Functions

We begin with an elementary review of the language of functions and, more importantly, the classification of functions according to purely set-theoretic criteria. We assume that the reader has seen the formal definition of a function elsewhere; in any case, we shall need nothing beyond the provisional definition implicit in the following paragraph.

1.1 Notation and Terminology

Let S and T be nonempty sets. Recall that a *function* (synonymously, a *map* or *mapping*) $f:S \to T$ is a pairing of elements from S and T such that each element $s \in S$ is associated with exactly one element $t \in T$, which is then denoted $f(s)$. If $f(s)=t$, we also say that t is *the image of s under f*. The set S is called the *domain* of f; the set T is called its *codomain*.

More conceptually, we may think of a function as consisting of three items of data: a domain S, a codomain T, and a rule of assignment f which maps every element of the domain to some element of the codomain. All three items participate in the definition, although often one simply speaks of the function f. Note that

(i) there is no requirement that every element of T occur as the image of some element in S;

(ii) it is possible that two distinct elements of S are assigned the same element of T.

In this sense, the definition of a function is highly asymmetric. We shall see below that the next two definitions, in tandem, restore the symmetry.

The set of all elements of T that occur as the image of some element in S is called the *range* or *image* of f and denoted $\mathrm{Im}(f)$. The following definitions are paramount.

DEFINITION. The function f is called *injective* or *one-to-one* if it satisfies the following condition:

$$\forall s, s' \in S, \quad f(s) = f(s') \Rightarrow s = s'$$

Equivalently,

$$\forall s, s' \in S, \quad s \neq s' \Rightarrow f(s) \neq f(s')$$

One may also say that f *separates points*.

DEFINITION. The function f is called *surjective* or *onto* if $\text{Im}(f) = T$; that is, if the following condition is satisfied:

$$\forall t \in T, \exists s \in S : f(s) = t$$

DEFINITION. The function f is called *bijective* if it is both injective and surjective.

EXAMPLES

(1) Let S be any nonempty set. Then the function

$$1_S : S \to S$$
$$s \mapsto s$$

is called the *identity map* on S. It is clearly bijective. (The notation $s \mapsto s$ indicates that 1_S has no effect on s.)

The next example shows clearly how the notions of injectivity and surjectivity depend not just on the rule of assignment, but essentially on the domain and codomain. First we recall the following standard elements of notation:

$\mathbf{R}$ = the set of real numbers
$\mathbf{R}_+$ = the set of nonnegative real numbers

(2) We give this example in four parts. First, the function

$$f : \mathbf{R} \to \mathbf{R}$$
$$x \mapsto x^2$$

is neither injective nor surjective. It is not injective since for any real number x, we have $f(x) = f(-x)$. It is not surjective, since no negative number occurs as the square of a real number. Second, the function

$$f: \mathbf{R} \to \mathbf{R}_+$$
$$x \mapsto x^2$$

is surjective, but not injective. The rule of assignment hasn't changed, but the codomain has, and every nonnegative real number is indeed the square of a real number. Third, the function

$$f: \mathbf{R}_+ \to \mathbf{R}$$
$$x \mapsto x^2$$

is injective, but not surjective: distinct nonnegative reals have distinct squares. Fourth, the function

$$f: \mathbf{R}_+ \to \mathbf{R}_+$$
$$x \mapsto x^2$$

is bijective. By restricting both the domain and the codomain we have eliminated both obstructions to bijectivity.

(3) Finally, consider differentiation as a function

$$\mathscr{C}^1(\mathbf{R}) \to \mathscr{C}^0(\mathbf{R})$$
$$f \mapsto \frac{df}{dx}$$

Here $\mathscr{C}^1(\mathbf{R})$ denotes the set of differentiable functions $f: \mathbf{R} \to \mathbf{R}$ with continuous derivative and $\mathscr{C}^0(\mathbf{R})$ denotes the set of merely continuous functions $f: \mathbf{R} \to \mathbf{R}$, with no assumption of differentiability. Viewed in this light, differentiation is not injective, since any two functions that differ by a constant have the same derivative, but it is surjective by the Fundamental Theorem of Calculus. For if f is any continuous real-valued function defined on $\mathbf{R}$, then for any fixed $a \in \mathbf{R}$,

$$\frac{d}{dx}\int_a^x f(t)\,dt = f(x)$$

which shows explicitly that f does indeed lie in the image of the differentiation operator.

1.2 Composition of Functions

Suppose we are given two functions $f: S \to T$ and $g: T \to U$. Then we can form their *composition* $g \circ f: S \to U$ defined by $g \circ f(s) = g(f(s))$. Composition is often represented by a *commutative diagram*:

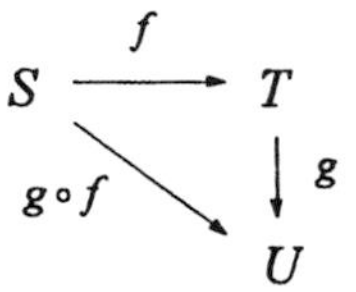

Figure 1.1. Commutative diagram.

This indicates that an element taking either path from S to U arrives at the same image.

Note that for any function $f: S \to T$ we have

$$f \circ 1_S = f \quad \text{and} \quad 1_T \circ f = f$$

so that the identity functions act neutrally with respect to composition.

1-1 PROPOSITION. *Let there be given functions*

$$f: S \to T, \quad g: T \to U, \quad h: U \to V$$

Then
$$h \circ (g \circ f) = (h \circ g) \circ f$$

Thus composition of functions, when defined, is associative.

PROOF. We show that both functions have the same effect on any element of S:

$$h \circ (g \circ f)(s) = h((g \circ f)(s)) = h(g(f(s))) = (h \circ g)(f(s)) = (h \circ g) \circ f(s)$$

This completes the proof. □

In view of this result, we need not write parentheses in connection with composition.

WARNING. Composition of functions is in general not defined in both directions, and when it is, it need not be commutative. For example, consider the functions $f(x)=x^2$ and $g(x)=x+1$ with domain and codomain assumed equal to the real numbers. Both $f \circ g$ and $g \circ f$ are defined, but clearly these functions are unequal.

1-2 PROPOSITION. *Let there be given functions* $f:S \to T$ *and* $g:T \to U$. *Then*

    (i) *if both f and g are injective, so is $g \circ f$;*

    (ii) *if both f and g are surjective, so is $g \circ f$;*

    (iii) *if both f and g are bijective, so is $g \circ f$;*

    (iv) *if $g \circ f$ is injective, then so is f;*

    (v) *if $g \circ f$ is surjective, then so is g.*

PROOF. (i) Suppose that $g \circ f(s)=g \circ f(s')$. We must show that the arguments are equal. We have the following chain of inferences:

$$g \circ f(s) = g \circ f(s') \implies g(f(s)) = g(f(s')) \qquad \text{by definition of composition}$$
$$\implies f(s) = f(s') \qquad \text{since } g \text{ is injective}$$
$$\implies s = s' \qquad \text{since } f \text{ is injective}$$

This establishes injectivity.

(ii) This is an embryonic example of a *diagram chase*. Referring to the commutative diagram illustrating composition of functions (Figure 1.1), given any element $u \in U$, we can lift it to an element $t \in T$ via the surjective map g. This element t can in turn be lifted to an element $s \in S$ via the surjective map f. By construction, s has image u under the composed map, which is therefore surjective.

(iii) This follows immediately from parts (i) and (ii).

(iv) We must show that two distinct arguments for f produce distinct images. Again we establish a chain of inferences:

$$s \neq s' \implies g \circ f(s) \neq g \circ f(s') \qquad \text{since } g \circ f \text{ is injective}$$
$$\implies g(f(s)) \neq g(f(s')) \qquad \text{by definition of composition}$$
$$\implies f(s) \neq f(s') \qquad \text{by definition of a function}$$

Thus f is injective, as claimed.

(v) This may be realized as another embryonic diagram chase, which we leave as an exercise in the style of (ii). Referring to the same picture, the student should work out the essentials and then carefully write out the argument.   ❑

1.3 Inverse Functions

DEFINITION. Let $f:S \to T$ be a function. Then f is called *invertible* if there exists a function $g:T \to S$ such that

$$g \circ f = 1_S \quad \text{and} \quad f \circ g = 1_T$$

In this case g is called an *inverse function* of f.

Note the symmetry of the definition: if g is an inverse function of f, then f is an inverse function of g.

1-3 PROPOSITION. *If f is invertible, its inverse is unique.*

PROOF. Suppose that f has inverse functions g and h. The following calculation shows that they are equal:

$$h = h \circ 1_T = h \circ (f \circ g) = (h \circ f) \circ g = 1_S \circ g = g \qquad ❑$$

In light of this result it is sensible to speak of *the* inverse of f, which is often denoted f^{-1}. By the earlier observation of symmetry, $(f^{-1})^{-1} = f$.

1-4 THEOREM. *A function $f:S \to T$ is invertible if and only if it is bijective.*

PROOF. $\Rightarrow$) Suppose that f is invertible with inverse function g. We shall show that it is both injective and surjective. By definition, $f \circ g = 1_T$; since the identity function is surjective, it follows from Proposition 1-2, part (v), that f must be surjective. Likewise, $g \circ f = 1_S$, which is injective, and hence f is injective.

$\Leftarrow$) Suppose that f is bijective. Then for each point $t \in T$, define $g(t)$ to be the unique $s \in S$ such that $f(s) = t$. (It exists by the surjectivity of f; it is unique by the injectivity.) Then clearly by construction,

$$g(f(s)) = s \quad \text{and} \quad f(g(t)) = t$$

This completes the proof.   ❑

EXAMPLE. Consider the exponential function

$$\mathbf{R} \to \mathbf{R}_+^{\times}$$
$$x \mapsto e^x$$

($\mathbf{R}_+^{\times}$ denotes the set of positive real numbers.) This is clearly bijective with well-known inverse

$$\mathbf{R}_+^{\times} \to \mathbf{R}$$
$$x \mapsto \ln(x)$$

We have the familiar formulas

$$e^{\ln(x)} = x \ (x \in \mathbf{R}_+^{\times}) \quad \text{and} \quad \ln(e^x) = x \ (x \in \mathbf{R})$$

as a special case of the definition of an inverse function.

1.4 Digression on Cardinality

This brief digression makes striking use of the concepts hitherto introduced, but is not used in any way in the sequel, except in the exercises for this chapter.

Let S and T be nonempty sets. Then one says that S and T are of the same *cardinality* and writes Card(S) = Card(T) if there exists a bijection $f: S \to T$. In the case of a finite set S having the same cardinality as the set $\{1,\ldots,n\}$, we abbreviate this to Card(S) = n. Note that this statement does not assert an equality of numbers, but rather affirms the existence of a bijection. Essentially cardinality generalizes the notion of counting to infinite sets.

One compares cardinalities in the following sense: If there exists at least an injective function $f: S \to T$, we write Card(S) $\leq$ Card(T). If there exists an injection, but no bijection, we write Card(S) < Card(T). This notation carries all of the usual properties of inequalities; in particular, we have the following celebrated theorem. (For the proof, see Exercise 21 below.)

THEOREM. (Schroeder-Bernstein) *If* Card(S) $\leq$ Card(T) *and* Card(T) $\leq$ Card(S), *then* Card(S) = Card(T).

Here are some amazing facts about cardinality. One can find proofs in any text on set theory or real analysis.

(1) Card(N) = Card(Z); that is, the set of natural numbers $N = \{0,1,2,...\}$ has the same cardinality as the set of integers $Z = \{...,-2,-1,0,+1,+2,...\}$.

(2) Card(Z) = Card(Q); that is, the set of integers has the same cardinality as the set of rational numbers (quotients of integers) Q.

(3) Card(Q) < Card(R); that is, the set of rational numbers has smaller cardinality than the set of real numbers R.

(4) Card(R) = Card(C); that is, the set of real numbers has the same cardinality as the set of complex numbers C.

(5) For any set S, let $\mathscr{P}(S)$ denote the set of all subsets of S. (This is called the *power set* of S.) Then if S is nonempty,

$$\text{Card}(S) < \text{Card}(\mathscr{P}(S))$$

Beginning with an infinite set S and iterating this operation, we can manufacture a never-ending ascending chain of infinite sets of strictly increasing cardinality. Hence there are infinitely many classes of infinity!

(6) Does there exist a set X such that Card(N) < Card(X) < Card(R)? This is unknown. The assertion that there is no such set is called the *continuum hypothesis*. (The question is somewhat more subtle than we have suggested here and is strongly tied to the very foundations of set theory and hence of mathematics.)

1.5 Permutations

Let $P_n = \{1,2,...,n\}$ denote the set consisting of the first n positive integers. Then the set of all bijective maps $P_n \to P_n$ is called *the symmetric group on n letters* and denoted S_n. (Note that for two finite sets of the same cardinality, the notions of injectivity, surjectivity, and bijectivity are equivalent; this is called the Pigeonhole Principle.) Elements of S_n are called *permutations*. We record some important properties of S_n, all of which follow from our previous work:

(i) Composition is an associative operation on S_n. In particular, S_n is closed under composition of functions.

(ii) The identity map in S_n acts as an identity element with respect to composition.

(iii) For every element f in S_n, there is an element g in S_n such that

$$f \circ g = g \circ f = 1_{P_n}$$

That is, every element of S_n has an inverse in S_n.

This last assertion is justified by the equivalence of bijectivity and invertibility of functions.

(One may well wonder why, of all the numerous properties of permutations, we have in particular cited these three. The answer will be made clear in the following chapter.)

EXAMPLE. Let $n=3$. We list the permutations by a 2×3 array which shows the image of each number directly below it.

$$\begin{pmatrix} 1 & 2 & 3 \\ 1 & 2 & 3 \end{pmatrix} \begin{pmatrix} 1 & 2 & 3 \\ 2 & 1 & 3 \end{pmatrix} \begin{pmatrix} 1 & 2 & 3 \\ 3 & 2 & 1 \end{pmatrix} \begin{pmatrix} 1 & 2 & 3 \\ 1 & 3 & 2 \end{pmatrix} \begin{pmatrix} 1 & 2 & 3 \\ 2 & 3 & 1 \end{pmatrix} \begin{pmatrix} 1 & 2 & 3 \\ 3 & 1 & 2 \end{pmatrix}$$

Note that this representation will work for any n.

1-5 PROPOSITION. *The cardinality of S_n is $n!$.*

PROOF. Consider the construction of a permutation: there are n choices for the image of 1, $n-1$ independent choices for the image of 2, $n-2$ independent choices for the image of 3, etc. Hence there are altogether $n \cdot (n-1) \cdots 3 \cdot 2 \cdot 1 = n!$ such permutations. ∎

We now pass to a more structural description of S_n. Let $a_1, a_2, \ldots, a_k$ be k distinct numbers in $\{1, \ldots, n\}$. Then the *k-cycle*

$$(a_1 \, a_2 \cdots a_k)$$

is the permutation defined by the following assignments:

$$a_1 \mapsto a_2, \ a_2 \mapsto a_3, \ldots, a_{k-1} \mapsto a_k, \ a_k \mapsto a_1$$

All other numbers are unaffected. In the special case of a 2-cycle, we speak of a *transposition*. We shall see shortly that all permutations may be constructed from transpositions.

EXAMPLES

(1) Write the following permutation in S_5 as a cycle.

$$\pi = \begin{pmatrix} 1 & 2 & 3 & 4 & 5 \\ 4 & 1 & 3 & 5 & 2 \end{pmatrix}$$

We see that 1 maps to 4, 4 to 5, 5 to 2, 2 back to 1. Accordingly

$$\pi = (1452)$$

(2) Express the following permutation in S_5 as a product of cycles.

$$\pi = \begin{pmatrix} 1 & 2 & 3 & 4 & 5 \\ 3 & 5 & 4 & 1 & 2 \end{pmatrix}$$

By tracing the *orbit* of each element, we find that

$$\pi = (134) \circ (25)$$

Note that these cycles have no elements in common.

(3) Express the following product of cycles in S_6 in row form.

$$\pi = (45) \circ (1364)$$

Recalling that composition of functions proceeds from right to left, we find that

$$\pi = \begin{pmatrix} 1 & 2 & 3 & 4 & 5 & 6 \\ 3 & 2 & 6 & 1 & 4 & 5 \end{pmatrix}$$

Generalizing our second example above, one can show that every permutation is expressible as the product (composition) of disjoint cycles. (Here disjointness means that no element occurs in more than one of the factors.) Moreover, every cycle can be written as the product of transpositions. For example,

$$(1234) = (14) \circ (13) \circ (12)$$

Thus every permutation can be factored into a product of transpositions. Now there is nothing unique about this factorization, but the following result, which is critical to linear algebra, does hold.

1-6 THEOREM. (*Invariance of Parity*) *Suppose that a permutation may be expressed as the product of an even number of transpositions. Then every factorization into transpositions likewise involves an even number of factors. Similarly, if a permutation may be expressed as the product of an odd number of transpositions, then every such factorization involves an odd number of transpositions.*

In other words, there may be different ways of expressing a permutation as a product of transpositions, but the *parity* of the number of factors is unique. Hence we may speak of a permutation as being either *odd* or *even* depending on whether it factors into an odd or even number of transpositions.

The proof of this theorem depends upon the construction of a powerful map

$$\sigma : S_n \to \{\pm 1\}$$

called the *sign homomorphism*. Its definition follows. (We shall fully explain the term *homomorphism* in Chapter 2.)

Let π lie in S_n. We say that π *reverses* the pair (i,j), if $i<j$, but $\pi(j)<\pi(i)$. It is easy to count the number of reversals when a permutation is expressed in matrix form: for every element in the second row, we count how many smaller elements lie to the right. For example, the permutation

$$\pi = \begin{pmatrix} 1 & 2 & 3 & 4 & 5 & 6 \\ 4 & 2 & 6 & 1 & 3 & 5 \end{pmatrix}$$

has $3+1+3+0+0+0=7$ reversals. Now if π has m reversals, define

$$\sigma(\pi) = (-1)^m$$

Hence the sign map is negative for permutations that have an odd number of reversals and positive for those that have an even number of reversals. It is clear that a transposition of adjacent elements, having exactly one reversal, has sign -1. The key result is this:

1-7 LEMMA. *Let π be a permutation and τ a transposition. Then*

$$\sigma(\tau \circ \pi) = -\sigma(\pi)$$

Thus composition with a transposition changes the sign of a permutation.

PROOF. This is easy to see in matrix form. Assume that $\pi \in S_n$ has the representation

$$\begin{pmatrix} 1 & 2 & \cdots & i & \cdots & j & \cdots & n \\ a_1 & a_2 & \cdots & a_i & \cdots & a_j & \cdots & a_n \end{pmatrix}$$

We ask what effect the transposition $(a_i a_j)$ has on π. To swap a_i and a_j we must first push a_i to the right across $j-i$ entries. (This amounts to $j-i$ adjacent transpositions.) Each move either adds or subtracts a reversal and hence changes the sign of the permutation once. We must next push a_j to the left across $j-i-1$ entries (one fewer), again changing the sign once for each move. In total we have made $2(j-i)-1$ sign changes. Since this number is manifestly odd, the sign of $\tau \circ \pi$ has indeed been changed relative to π, as claimed.   □

By repeated use of the lemma it follows that for the product of m transpositions

$$\sigma(\tau_1 \circ \tau_2 \circ \cdots \circ \tau_m) = (-1)^m$$

Now the proof of Theorem 1-6 is clear: Suppose we have two equal products of transpositions

$$\tau_1 \circ \tau_2 \circ \cdots \circ \tau_m = \omega_1 \circ \omega_2 \circ \cdots \circ \omega_{m'}$$

Then applying σ to both sides, we find that

$$(-1)^m = (-1)^{m'}$$

and therefore m and m' have the same parity, as claimed.

Exercises

1. Find sets S, T, and U and functions $f: S \rightarrow T$ and $g: T \rightarrow U$ such that $g \circ f$ is injective, but g is *not* injective. (*Hint*: The choice of $S=\{0\}$, $T=\{0,1\}$, and $U=\{0\}$ will inevitably lead to the desired result.)

2. Find sets S, T, and U and functions $f: S \rightarrow T$ and $g: T \rightarrow U$ such that $g \circ f$ is surjective, but f is *not* surjective.

3. Find a non-identity function $\mathbf{R} \rightarrow \mathbf{R}$ which is its own inverse function. That is, $f \circ f = 1_{\mathbf{R}}$ or, equivalently, $f(f(x))=x$ for all real x.

4. Assess the injectivity and surjectivity of each of the following functions from $\mathbf{R}$ to $\mathbf{R}$. Be sure to defend your responses rigorously. For example, to show that a function is bijective, you might exhibit the inverse function.

You might also deduce valuable information by sketching the graph as you did in calculus.

(a) $f(x) = 2x + 1$

(b) $f(x) = x^3 - 3x$

(c) $f(x) = e^x$

(d) $f(x) = x^4 - 2x^2$

(e) $f(x) = e^{-x^2}$

5. Find a bijective map from the open interval $(-\pi/2, +\pi/2)$ to the set **R** of real numbers. This shows, by the way, that both sets have the same cardinality. (*Hint*: Think trigonometrically.)

6. Explicitly construct a bijective function from the set of integers **Z** to the set of even integers 2**Z**. Despite your naive intuition to the contrary, this shows that Card(**Z**) = Card(2**Z**).

7. Prove that if Card(S) = Card(T) and Card(T) = Card(U), then Card(S) = Card(U). (*Hint*: Unwind the definition of equality of cardinality and check your basic facts about composition of functions.)

8. Let N×N denote the set of all ordered pairs (m,n) of natural numbers. [For example, (1,2), (0,7), etc.] Show that Card(N×N) = Card(N) by explicitly constructing a bijection N×N→N. Be sure to explain your answer carefully. (*Hint*: Plot a few of the points of N×N on an ordinary *xy*-graph. Look for a path through the nodes which visits every point of N×N exactly once. This allows you to enumerate the points—in other words, to construct a bijection with the set N. In fact, it is not hard to write down an explicit formula.)

9. Express the following composition of cycles in matrix notation.

$$(1\ 4\ 7\ 6) \circ (2\ 7\ 5)$$

Be sure to work from right to left.

10. Decompose the following permutation into the product of disjoint cycles. (Recall that disjointness means that no element occurs in more than one of the cycles.)

$$\pi = \begin{pmatrix} 1 & 2 & 3 & 4 & 5 & 6 & 7 & 8 & 9 \\ 6 & 5 & 9 & 1 & 4 & 3 & 8 & 7 & 2 \end{pmatrix}$$

11. How many reversals does the permutation π in the previous problem manifest? What is its sign? Factor π into the product of transpositions.

12. Referring to the permutation,

$$\pi = \begin{pmatrix} 1 & 2 & 3 & 4 & 5 & 6 & 7 & 8 \\ 8 & 1 & 3 & 6 & 4 & 5 & 7 & 2 \end{pmatrix}$$

   (a) express π as the product of disjoint cycles;

   (b) express π as the product of transpositions;

   (c) compute the sign of π.

13. Let $\pi \in S_n$ be a k-cycle. For which positive integers n is π^n equal to the identity map? Here π^n means π composed with itself n times. (*Hint*: Try some examples for small k.)

14. Suppose that for permutations π and τ in S_n, $\pi \circ \tau = \tau \circ \pi$. (Recall that this is not in general true.) Show that for all positive integers n, $(\pi \circ \tau)^n = \pi^n \circ \tau^n$.

15. Show that disjoint cycles commute; that is, if π and τ are cycles with no elements in common, then $\pi \circ \tau = \tau \circ \pi$.

16. Recall that every permutation may be written as the product of disjoint cycles. Using this fact and the previous three exercises, show that $\pi^{n!}$ is the identity map for any permutation π in S_n. (This says, in particular, that if one shuffles a deck of 52 cards in exactly the same way 52! times, the deck will return to its original state. Since 52! is approximately 8×10^{67}, it is far easier to prove this result than to establish it empirically.)

17. We showed above that for transpositions $\tau_1, \tau_2, \ldots, \tau_m$

$$\sigma(\tau_1 \circ \tau_2 \circ \ldots \circ \tau_m) = (-1)^m$$

and also that every permutation is the product of transpositions. Use these two facts to prove that for all permutations π_1 and π_2,

$$\sigma(\pi_1 \circ \pi_2) = \sigma(\pi_1)\sigma(\pi_2)$$

We shall see shortly in Section 2.2 that this special property explains why the sign map $\sigma: S_n \rightarrow \{\pm 1\}$ is called a *homomorphism*.

Let S and T be nonempty sets. Recall that their *Cartesian product* $S \times T$ consists of all ordered pairs (s,t) such that $s \in S$ and $t \in T$. By *ordered pair* we mean that

$$(s,t) = (s',t') \Leftrightarrow s = s' \text{ and } t = t'$$

(For a rigorous set-theoretic definition of this concept, see any book on set theory.)

18. Show that if S and T are nonempty sets, then Card($S \times T$) = Card($T \times S$). [*Hint*: Explicitly construct the bijection. Where in $T \times S$ can you send (s,t)? What is the inverse map?]

Recall that sets S and T are called *disjoint* if they have no common elements; that is, if $S \cap T = \varnothing$. In this case, $S \cup T$ is referred to as the *disjoint union* of S and T.

19. Let S, S', T, T' be sets such that S and T are disjoint, as are S' and T'. Assume further that Card(S) = Card(S') and Card(T) = Card(T'). Show that Card($S \cup T$) = Card($S' \cup T'$). Note that this result extends easily to arbitrary disjoint unions (possibly involving an infinite family of disjoint sets).

20. Let S and T be nonempty sets. Show that there exist sets S' and T' such that Card(S) = Card(S') and Card(T) = Card(T') with S' and T' disjoint. (*Hint*: Take $S' = S \times \{0\}$ and $T' = T \times \{1\}$. Can S' and T' have any elements in common?)

The final problem constitutes a short, elegant proof of the Schroeder-Bernstein Theorem. We assume that S and T are nonempty sets and that there exist injective maps $f: S \rightarrow T$ and $g: T \rightarrow S$. This amounts to the assertion that both Card(S) $\leq$ Card(T) and Card(T) $\leq$ Card(S). Since we are only interested in cardinalities, by the previous exercise we may assume that S and T are disjoint.

21. Given any $u \in S$, define a sequence

$$\Gamma_u = \{\dots, s_{-2}, t_{-2}, s_{-1}, t_{-1}, s_0, t_0, s_1, t_1, s_2, t_2, \dots\}$$

in $S \cup T$ as follows. For positive indices,

$$s_0 = u$$
$$t_0 = f(s_0)$$
$$s_1 = g(t_0)$$
$$t_1 = f(s_1)$$
$$s_2 = g(t_1) \text{ etc.}$$

Hence the sequence goes on forever in this direction. For negative indices,

$$t_{-1} = g^{-1}(s_0)$$
$$s_{-1} = f^{-1}(t_{-1})$$
$$t_{-2} = g^{-1}(s_{-1})$$
$$s_{-2} = f^{-1}(t_{-2}) \text{ etc.}$$

provided that these inverse images exist! Otherwise, the sequence terminates. (We are really abusing the notation here, since the inverse functions f^{-1} and g^{-1} are not assumed to exist. But whenever the inverse image of a given element does exist, it is unique since both f and g are assumed injective.) Note that there are three possibilities:

(i)   the sequence continues to the left without limit;

(ii)   the sequence terminates on the left at an element of S;

(iii) the sequence terminates on the left at an element of T.

Now define $S_u \subseteq S$ to be those elements of S that occur in the sequence Γ_u and similarly define $T_u \subseteq T$ to be those elements of T that occur in Γ_u.

(a) Show that for any pair of elements $u, v \in S$, either $S_u = S_v$ or S_u and S_v are disjoint; show that an analogous statement holds for T_u and T_v. Since for all $u \in S$ and $t \in T$, we have $u \in S_u$ and $t \in T_{g(t)}$ (why?), this implies that S is the disjoint union of subsets of the form S_u and similarly for T.

(b) Show that for each $u \in S$, $\text{Card}(S_u) = \text{Card}(T_u)$. (*Hint*: Consider each of the three possibilities for Γ_u introduced above. Doesn't each suggest an obvious map from S_u to T_u?)

(c) Show that Card(S) = Card(T). This completes the proof of the Schroeder-Bernstein Theorem. [*Hint*: Use part (b) and Exercise 19.]

The argument outlined in this exercise is adapted from Paul J. Cohen's classic monograph *Set Theory and the Continuum Hypothesis*.

2
Groups and Group Homomorphisms

This chapter introduces the notion of an abstract group, one of the fundamental objects of both algebra and geometry and very much at the heart of linear algebra. In the early stages, the student will perhaps see only a rather arbitrary looking (but attractive!) informal axiomatic system. This is a gross deception. The definition distills millennia of mathematical experience. Another theme also emerges: objects are not nearly so interesting in themselves as the relationships they bear to one another. In the case of groups, these relationships are expressed by group homomorphisms.

2.1 Groups and Subgroups

Let S be a nonempty set. Recall that $S{\times}S$ denotes the set of all ordered pairs (s,t) of elements in S. A *binary operation* on S is a function $S{\times}S \to S$. We almost invariably denote the image of (s,t) under this operation with some infix operator such as $*$. Hence

$$(s,t) \mapsto s*t$$

Note that the notion of *closure* is implicit in our definition since the codomain of the operation is S. Often, however, we wish to pass from a set S to a subset T and consider the status of $*$ as an operation on T. In this case, closure is not at all implicit and must be verified.

An operation is called *associative* if

$$(s*t)*u = s*(t*u)$$

for all $s,t,u \in S$. We say that $e \in S$ is an *identity* for $*$ if

$$s*e = s = e*s$$

for all $s \in S$. (In general, an identity need not exist.) Finally, we say that $*$ is *commutative* if

$$s*t = t*s$$

for all $s,t \in S$.

DEFINITIONS. A *magma* $\langle S,* \rangle$ is a set S together with an operation $*$ (and no other assumptions). An associative magma is called a *semigroup*. A semigroup with an identity element is called a *monoid*. Each of these objects is called *commutative* if the operation in question is commutative.

With one further requirement, we reach one of the fundamental structures in all of mathematics.

DEFINITION. A *group* $\langle G,* \rangle$ is a monoid in which every element is invertible. Specifically,

(i)   the operation $*$ is associative on G;

(ii) there exists an element $e \in G$ which is an identity for $*$;

(iii) for every $s \in G$ there exists a $t \in G$ such that $s*t = e = t*s$.

With regard to this last condition, we say that t is an *inverse* for s. By symmetry, s is likewise an inverse for t.

A group is called *commutative* if the corresponding operation is. A commutative group with operation denoted $+$ is called an *additive group*. The identity of an additive group is usually denoted 0. Note that this may not be the familiar zero of ordinary arithmetic.

EXAMPLES

The following magmas are largely familiar objects. Nonetheless, the reader should explicitly verify all applicable axioms.

(1) $\langle \mathbf{N},+ \rangle$, the set of natural numbers under addition, is a commutative monoid (with identity 0), but not a group, since only 0 has an inverse (itself).

(2) $\langle \mathbf{Z},+ \rangle$, the set of integers under addition, is an additive group. The additive inverse of n is $-n$.

(3) $\langle \mathbf{Q},* \rangle$, the set of rational numbers under multiplication, is a commutative monoid, but not a group, since 0 is not invertible.

(4) $\langle \mathbf{Q}^*,* \rangle$, the set of nonzero rational numbers under multiplication, is a commutative group.

(5) The set $\{-1,+1\}$ forms a commutative finite group with respect to multiplication. For those familiar with complex numbers, so does $\{+i,-1,-i,+1\}$.

(6) For any positive n, $\langle S_n,\circ \rangle$ constitutes a finite group of order $n!$; this is precisely the content of the analysis given at the start of Section 1.5. One can show easily that S_n is noncommutative for $n > 2$.

(7) The set $\mathscr{C}^0(\mathbf{R})$ of continuous real-valued functions defined on $\mathbf{R}$ forms an additive group with respect to addition of functions. (Closure is verified by noting that the sum of two continuous functions is again continuous.) The additive inverse of $f(x)$ is $-f(x)$.

Observe that $\mathscr{C}^0(\mathbf{R})$ constitutes no more than a monoid with respect to composition of functions. Why?

(8) Here is a small additive group of 5 elements, often denoted Z_5.

+	0	1	2	3	4
0	0	1	2	3	4
1	1	2	3	4	0
2	2	3	4	0	1
3	3	4	0	1	2
4	4	0	1	2	3

There is little mystery to the operation: we simply add the operands as usual and then discard all but the remainder when divided by 5. This construction can be adapted to any positive integer n to form $\mathbf{Z}_n$, the additive group of integers modulo n. Clearly Card$(\mathbf{Z}_n)=n$.

The explicit listing of products in the last example is called a *Cayley table*. The name honors Arthur Cayley (1821–95), who first introduced the notion of an abstract group in the middle of the nineteenth century. Curiously enough, the definition was far ahead of its time and ignored for many decades.

Unless we require special emphasis on the operator, in the sequel we shall usually write st for the product $s*t$, except in the additive case, for which we always write $s+t$. Also, in view of the associative law, we shall often drop the parentheses for products of three or more group elements.

The following propositions summarize some of the most basic properties of groups. These facts, together with their proofs, should become part of the student's psyche.

2-1 PROPOSITION. (Cancellation Laws) *Let G be a group. Then for any three elements s,t,u∈G,*

$$st = su \Rightarrow t = u$$
$$st = ut \Rightarrow s = u$$

PROOF. Suppose that $st=su$. Then since s is invertible, there exists an element x in G, such that $xs = e$. Using the associative law and the definition of an identity, we now have the following chain of equalities:

$$st = su$$
$$x(st) = x(su)$$
$$(xs)t = (xs)u$$
$$et = eu$$
$$t = u$$

The second assertion is proved similarly. ◻

In additive notation, this result becomes

$$s+t = s+u \Rightarrow t = u$$
$$s+t = u+t \Rightarrow s = u$$

2-2 PROPOSITION. *Let G be a group. Then the following assertions hold:*

(i) *The identity element e∈G is unique.*

(ii) *Inverses are unique; that is, for every s∈G there exists a unique t∈G such that st=e=ts. Henceforth we denote this unique inverse s^{-1}.*

(iii) *If s,t∈G with st=e, then $s=t^{-1}$ and $t=s^{-1}$. Hence to check inverses, we need only check on one side.*

(iv) *For all s∈G, $(s^{-1})^{-1}=s$.*

(v) *For all s,t∈G, $(st)^{-1}=t^{-1}s^{-1}$.*

(vi) *If s∈G, then ss=s if and only if s=e.*

PROOF. (i) Suppose that both e and e' are identities. Then by the definition of an identity for the group operation, we have $e = ee' = e'$, which proves uniqueness. Note that this argument works in any magma.

(ii) Suppose that both t and t' are inverses for $s \in G$. Then $st = e = st'$, and by left cancellation we have $t = t'$.

(iii) Suppose that $st = e$. Multiply both sides by s^{-1} to find that

$$s^{-1}(st) = s^{-1}$$

which via associativity shows at once that $s^{-1} = t$.

(iv) We have already observed that if t is an inverse for s, then s is an inverse for t. But in view of the uniqueness of inverses, this is precisely the content of the assertion.

(v) We compute the product

$$(st)(t^{-1}s^{-1}) = s(tt^{-1})s^{-1} = ses^{-1} = ss^{-1} = e$$

According to part (iii) above, this suffices to show that $t^{-1}s^{-1} = (st)^{-1}$, as claimed.

(vi) Clearly $ee = e$. If $ss = s$, we cancel an s from each side of the equation to find that $s = e$. Thus e is the only element satisfying this property. $\square$

An element s of a magma such that $ss = s$ is called *idempotent*. Hence part (vi) of the proposition states that the identity is the only idempotent element of a group.

Note that we can define exponentials in groups as we do in ordinary algebra:

$$s^n = s \cdot s \cdots s \qquad (n \text{ times})$$
$$s^{-n} = s^{-1} \cdot s^{-1} \cdots s^{-1} \quad (n \text{ times})$$

for positive n, and s^0 is defined to be the identity e. Take care, however! The familiar formula

$$(st)^n = s^n t^n$$

holds for all n if and only if we are working in a commutative group. That this is true for a commutative group is obvious. To see the converse, note that on the one hand,

$$(st)^2 = stst$$

while on the other,

$$s^2t^2 = sstt$$

If these expressions are equal, $sstt = stst$ and we can cancel s from the left and t from the right to obtain $st = ts$.

In additive notation, the inverse of s in G is denoted $-s$ and we abbreviate the expression $s + (-t)$ to $s - t$. Repeated addition is expressed via integral coefficients. Thus

$$ns = s + s + \cdots + s \quad (n \text{ times})$$

Interpreting the previous proposition and our discussion of exponents in this special case, we have many familiar arithmetic properties, such as these:

$$-(-s) = s$$
$$-(s+t) = -s - t$$
$$n(s+t) = ns + nt$$

for all $s, t \in G$ and integers n. Note that the last two identities are entirely dependent on commutativity.

Subgroups

DEFINITION. Let $\langle G, * \rangle$ be a group. Then a subset H of G is called a *subgroup* of G if it constitutes a group in its own right with respect to the operation $*$ defined on G.

We can always show directly that a nonempty subset H of G is a subgroup by verifying that

(i) H is closed under the operation defined on G.

(ii) H contains the identity of G. (Why can't there be a "restricted" identity— one that works in H but is not the identity of the full group?)

(iii) If s lies in H, then so does s^{-1}.

The point is that associativity is inherited from the ambient group.

2-3 PROPOSITION. *A nonempty subset H of a group G is a subgroup of G if and only if the following condition holds:*

$$s, t \in H \Rightarrow st^{-1} \in H$$

Thus the problem of checking that a nonempty subset is a subgroup is often reduced to one step.

PROOF. $\Rightarrow$) Let H be a subgroup and suppose that s and t lie in H. Then t^{-1} lies in H, and therefore so does the product st^{-1} by closure.

$\Leftarrow$) Suppose that H satisfies the condition stated in the proposition. We shall verify each of the points (i), (ii), and (iii) above. First (ii). Since H is nonempty, it contains some element s. But then by assumption, H also contains the product $ss^{-1}=e$, as required. Now since H contains s and e, it also contains $es^{-1}=s^{-1}$, which verifies condition (iii). Finally, if s and t lie in H, then by the previous step, so does t^{-1}. But again by hypothesis this implies that the product $s(t^{-1})^{-1}=st$ lies in H, thus establishing closure. $\square$

In additive notation, the condition of the proposition is

$$s,t \in H \Rightarrow s-t \in H$$

Note that a group G is always a subgroup of itself. Other subgroups are called *proper subgroups*. Likewise, the subset $\{e\}$ always constitutes a subgroup of G, called the *trivial subgroup*. The student should now find, if possible, nontrivial proper subgroups for each of the examples of groups listed above. We give a few of these.

EXAMPLES

(1) For any integer n, define

$$n\mathbf{Z} = \{na : a \in \mathbf{Z}\}$$

(i.e., all multiples of n). Then $n\mathbf{Z}$ is a subgroup of the additive group $\mathbf{Z}$. This is clear by Proposition 2-3 since $na-nb=n(a-b)$, whence the difference of two elements chosen from $n\mathbf{Z}$ again lies therein. Note that this subgroup is nontrivial if n is not 0 and proper if n is not ±1.

(2) For any nonzero rational number a,

$$\{a^n : n \in \mathbf{Z}\}$$

(i.e., all integral powers of a) is a subgroup of the multiplicative group $\mathbf{Q}^*$. Check this as in Example 1.

(3) Consider the additive group $\mathscr{C}^0(\mathbf{R})$ of continuous real-valued functions defined on $\mathbf{R}$. Let I be the set of all functions whose value at 0 is 0. Then I is

a subgroup of $\mathscr{C}^0(\mathbf{R})$. For if f and g lie in I, then

$$(f-g)(0) = f(0)-g(0) = 0-0 = 0$$

whence the difference is also in I. This example generalizes readily.

(4) The set $\{\bar{0},\bar{2},\bar{4}\}$ is a subgroup of the additive group $\mathbf{Z}_6$ defined above. (Here we use the bar to distinguish these elements from ordinary integers.) One easily verifies this directly.

2.2 Group Homomorphisms

DEFINITION. Let G_0 and G_1 be groups. Then a function $\varphi: G_0 \to G_1$ is called a *homomorphism of groups* if it satisfies the following condition:

$$\varphi(st) = \varphi(s)\varphi(t)$$

for all $s,t \in G_0$.

Note carefully that the implied operations occur in (possibly) different groups, the first in the domain, the second in the codomain.

In additive notation, the condition that defines a group homomorphism amounts to

$$\varphi(s+t) = \varphi(s)+\varphi(t)$$

for all $s,t \in G_0$.

EXAMPLES

(1) Let a be any real number. Then the function $f:\mathbf{R} \to \mathbf{R}$ defined by $f(x)=ax$ is a homomorphism from the group of real numbers under ordinary addition to itself. We have only to check that

$$f(x+y) = a(x+y) = ax + ay = f(x) + f(y)$$

(2) Differentiation is a group homomorphism $\mathscr{C}^1(\mathbf{R}) \to \mathscr{C}^0(\mathbf{R})$. This is essentially the content of the familiar rule of differentiation

$$\frac{d}{dx}(f+g) = \frac{df}{dx} + \frac{dg}{dx}$$

(3) The exponential map

$$\mathbf{R} \to \mathbf{R}_+^\times$$
$$x \mapsto e^x$$

(regarded by some as the most important function in mathematics) is a homomorphism from the group of real numbers under addition to the group of positive real numbers under multiplication. To see this, note that

$$e^{x+y} = e^x e^y$$

(4) The sign homomorphism $\sigma: S_n \to \{\pm 1\}$ is a group homomorphism from the symmetric group on n letters (under composition) to the multiplicative group $\{\pm 1\}$. The requisite property that

$$\sigma(\pi_1 \pi_2) = \sigma(\pi_1)\sigma(\pi_2)$$

was shown in Chapter 1, Exercise 17.

Whenever we introduce a class of maps that respect the structure of both domain and codomain in the sense that group homomorphisms do, a proposition of the following form is in order.

2-4 PROPOSITION. *The composition of group homomorphisms is a group homomorphism.*

PROOF. Let $\varphi_0: G_0 \to G_1$ and $\varphi_1: G_1 \to G_2$ be group homomorphisms. Then for all $s, t \in G_0$, we have

$$\varphi_1 \circ \varphi_0(st) = \varphi_1(\varphi_0(st)) = \varphi_1(\varphi_0(s)\varphi_0(t)) = \varphi_1(\varphi_0(s))\varphi_1(\varphi_0(t)) = \varphi_1 \circ \varphi_0(s)\varphi_1 \circ \varphi_0(t)$$

This completes the proof.   □

2-5 PROPOSITION. *Let $\varphi: G_0 \to G_1$ be a homomorphism of groups. Then the following assertions hold:*

(i) $\varphi(e_0) = e_1$, *where e_j is the identity of G_j, $j = 0, 1$.*

(ii) $\varphi(s^{-1}) = \varphi(s)^{-1}$ *for all $s \in G_0$.*

(iii) $\varphi(s^m) = \varphi(s)^m$ *for all $s \in G_0$ and integers m.*

In additive notation (in which form we shall most often meet this result), these properties read as follows:

(i) $\varphi(0)=0$

(ii) $\varphi(-s)=-\varphi(s)$

(iii) $\varphi(ms)=m\varphi(s)$

PROOF. (i) By the definition of a homomorphism and an identity

$$\varphi(e_0) = \varphi(e_0 e_0) = \varphi(e_0)\varphi(e_0)$$

whence $\varphi(e_0)$ is idempotent (i.e., equal to its product with itself). But the only element in G_1 with this property is e_1 by part (vi) of Proposition 2-2.

(ii) For all $s \in G_0$,

$$\varphi(s)\varphi(s^{-1}) = \varphi(ss^{-1}) = \varphi(e_0) = e_1$$

hence $\varphi(s^{-1})=\varphi(s)^{-1}$ by part (iii) of the proposition just cited.

(iii) Exercise. [*Hint*: Distinguish between the cases of positive and negative integers m and use part (ii).] $\square$

A bijective homomorphism of groups $\varphi: G_0 \to G_1$ is called an *isomorphism of groups* or a *group isomorphism*. One writes $G_0 \cong G_1$ to indicate the existence of an isomorphism. This says that the groups are structurally identical. So, for instance, Example 3 above shows that

$$\langle \mathbf{R},+\rangle \cong \langle \mathbf{R}_+^\times,*\rangle$$

which is remarkable. We leave it as an exercise to show that if φ is an isomorphism, then so is the inverse map φ^{-1}.

Before proceeding with the theory of group homomorphisms, we interject a purely set-theoretic notion. If $f: S \to T$ is a function, then for any $t \in T$, we define the *inverse image of t under f*, henceforth denoted $f^{-1}(t)$, to be the set of all $s \in S$ such that $f(s) = t$. Note that the inverse image is empty if t does not lie in the image of f and may contain more than one element if f is not injective. Hence the notation $f^{-1}(t)$ makes sense even when f fails to be invertible as defined in Chapter 1. (This is a slight abuse of notation: the inverse image of t under f in the sense just introduced is technically a subset of S; the value of the inverse function—should one exist—is, for any given $t \in T$, just a point $s \in S$.)

DEFINITION. Let $\varphi: G_0 \to G_1$ be a group homomorphism. Then the *kernel* of φ, denoted Ker(φ), is defined by

$$\text{Ker}(\varphi) = \{s \in G_0 : \varphi(s)=e\}$$

That is, $\text{Ker}(\varphi) = \varphi^{-1}(e)$, the inverse image of the identity of the codomain. (Note that we are no longer subscripting the identity; its meaning is implicit in the context.) We shall see shortly that the kernel of a homomorphism reflects the amount of information lost in applying the map.

The *image* of φ, $\text{Im}(\varphi)$, is its image in the ordinary function-theoretic sense; that is,

$$\text{Im}(\varphi) = \{t \in G_1 : \exists s \in G_0 \text{ such that } \varphi(s) = t\}$$

A result of the following type should be no surprise for structure-preserving maps.

2-6 PROPOSITION. *Both the image and the kernel of a group homomorphism are subgroups of their ambient groups.*

PROOF. We show that the kernel is a subgroup; the image is left to the student. First note that the kernel is never empty since it always contains the identity. Let s and t lie in the kernel of a homomorphism $\varphi: G_0 \to G_1$. Then by the definition and elementary properties of homomorphisms, we have

$$\varphi(st^{-1}) = \varphi(s)\varphi(t^{-1}) = \varphi(s)\varphi(t)^{-1} = ee^{-1} = e$$

Hence the product st^{-1} is also in the kernel, and by Proposition 2-3 this suffices to establish that the kernel is a subgroup. ☐

Finally we examine one of the most important properties of a group homomorphism: the efficient characterization of inverse images.

2-7 PROPOSITION. *Let $\varphi: G_0 \to G_1$ be a homomorphism of groups and suppose that $\varphi(s) = t$. Then*

$$\varphi^{-1}(t) = \{sk : k \in \text{Ker}(\varphi)\}$$

That is, the complete solution set to the equation $\varphi(x) = t$ is precisely the set of all products sk where s is any particular solution and k lies in the kernel of φ.

In additive notation, this reads

$$\varphi^{-1}(t) = \{s + k : k \in \text{Ker}(\varphi)\}$$

which is to say that the solution set to the equation $\varphi(x) = t$ consists of all sums of the form $s+k$, where s is any particular solution and k ranges over the kernel of the map φ.

PROOF. We have two things to show: one, that every element of the form sk with $k \in \mathrm{Ker}(\varphi)$ lies in the inverse image of t under φ; two, that every element in the inverse image of t is indeed of this form.

Part One. Consider the product sk where k lies in the kernel. We have

$$\varphi(sk) = \varphi(s)\varphi(k) = te = t$$

since φ is a homomorphism which by assumption sends s to t, and since furthermore elements in the kernel by definition map onto the identity e.

Part Two. Suppose that u also lies in the inverse image of t. Then by assumption

$$\varphi(u) = \varphi(s)$$

and we have

$$\varphi(s^{-1}u) = \varphi(s^{-1})\varphi(u) = \varphi(s)^{-1}\varphi(u) = t^{-1}t = e$$

Therefore the product $s^{-1}u$ is equal to some element k of the kernel. But clearly if $s^{-1}u=k$, then $u=sk$, and u has the required form. This concludes the proof. $\square$

2-8 COROLLARY. *A homomorphism of groups is injective if and only if it has trivial kernel (i.e., its kernel consists only of the identity element).*

PROOF. If a homomorphism φ is injective, the inverse image of the identity of the codomain can only contain one object—the identity of the domain. Conversely, according to Proposition 2-7, the complete inverse image of any element t of the codomain is either empty or consists of products of the form sk where k is in the kernel and s has image t. But if the kernel is trivial, there is only one such product, $se=s$. Hence there is at most one element in the inverse image of t, which is precisely to say that φ is injective. $\square$

EXAMPLES

(1) Consider the mapping of multiplicative groups

$$\mathbf{R}^* \to \mathbf{R}^*$$

$$x \mapsto x^2$$

This is a homomorphism—albeit a dreary one—since $(xy)^2=x^2y^2$. The kernel of this noninjective map is $\{\pm 1\}$. What is the inverse image of a nonnegative real number y? In accordance with Proposition 2-7, it is $\pm y^{1/2}$. The kernel thus reflects the loss of the sign in squaring a real number.

(2) Reconsider differentiation, which we now regard as an additive group homomorphism $D:\mathscr{C}^1(\mathbf{R})\rightarrow\mathscr{C}^0(\mathbf{R})$ with $D(f)=df/dx$. We know from calculus that

$$\mathrm{Ker}(D) = \{f:f(x)=c \text{ for all } x, \text{ for some constant } c\in\mathbf{R}\}$$

That is, the kernel consists of all constant functions and thus reflects the loss of constant terms under differentiation. Now the interpretation of our last proposition is quite familiar: if $F\in\mathscr{C}^1(\mathbf{R})$ is such that $D(F)=f$, then

$$D^{-1}(f) = \{F+c : c\in\mathbf{R}\}$$

This is perhaps better recognized in its more customary form:

$$\int f(x)dx = F(x)+c$$

where F is any antiderivative of f (i.e., a particular solution to $dF/dx=f$) and c is an arbitrary real constant [i.e., an element of $\mathrm{Ker}(D)$].

2.3 Rings and Fields

We have seen that a group is an extremely general algebraic structure which admits a vast range of specific instances. We now briefly explore a specialization with quite a different flavor. We shall be concerned simultaneously with two operations and an essential property that intertwines them.

DEFINITION. A *ring* (with unity) consists of a nonempty set A together with two operations $+$ and $*$ such that the following properties hold:

(i) $\langle A,+\rangle$ is an additive group.

(ii) $\langle A,*\rangle$ is a monoid (i.e., an associative magma with an identity, which in this case we shall always denote as 1).

(iii) These operations satisfy two *distributive laws*, asserting that

$$a*(b+c) = a*b + a*c \quad \text{and} \quad (a+b)*c = a*c + b*c$$

for any three elements $a,b,c\in A$.

Note that the operation * is not necessarily commutative. If it is, we speak of a *commutative ring*. Ordinarily we shall refer to + as addition and * as multiplication, even though these terms may not carry their usual meaning. Again it is customary to write ab for $a*b$.

Before giving examples, we note one critical special case.

DEFINITION. A commutative ring k is called a *field* if k^*, the set of nonzero elements of k, forms a group with respect to the ring multiplication. Thus every nonzero element of k has a multiplicative inverse.

EXAMPLES

(1) The integers **Z** constitute a commutative ring with respect to ordinary addition and multiplication. However, **Z** is not a field because $+1$ and -1 are its only invertible elements with respect to multiplication.

(2) The rational numbers **Q** constitute a field with respect to ordinary addition and multiplication. So do **R**, the real numbers, and **C**, the complex numbers.

(3) The set $\mathscr{C}^0(\mathbf{R})$ of continuous real-valued functions with domain **R** is a commutative ring, but not a field, with respect to addition and multiplication of functions. It inherits the requisite properties from the corresponding properties of real numbers. The identity with respect to multiplication is the constant function $f(x)=1$.

(4) Let **Z**[x] denote the set of all polynomials in the indeterminate x with integral coefficients. Then **Z**[x] is a commutative ring with respect to addition and multiplication of polynomials; the required properties are all familiar laws of elementary algebra. Similarly, we can form the commutative rings **Q**[x], **R**[x], and **C**[x]. In fact, if k is any commutative ring whatsoever, one can define the polynomial ring $k[x]$, although its interpretation as a collection of functions must be carefully reviewed.

(5) The following pair of Cayley tables defines a field structure on a set consisting of only three elements. This field is denoted $\mathbf{F}_3$. Note that we have merely added a multiplicative structure to $\mathbf{Z}_3$.

+	0	1	2
0	0	1	2
1	1	2	0
2	2	0	1

*	0	1	2
0	0	0	0
1	0	1	2
2	0	2	1

The tables are computed as in Example 8 of Section 2.1 above: add or multiply as ordinary integers and then discard all but the remainder of division by 3.

This construction generalizes to any prime number p to yield the finite field $\mathbf{F}_p$. What happens if we do not use a prime modulus? The student can show that we still obtain a commutative ring, but never a field. (See Exercises 21–25 at the end of this chapter.)

The following proposition summarizes those arithmetic properties of rings which are used routinely. All of the properties of additive groups, of course, remain in effect.

2-9 PROPOSITION. *Let A be a ring with unity. Then the following assertions hold:*

(i) $0a=0=a0, \ \forall a \in A$

(ii) $a(-b)=-(ab)=(-a)b, \ \forall a,b \in A$

(iii) $(-a)(-b)=ab, \ \forall a,b \in A$

(iv) $(-1)a=-a, \ \forall a \in A$

(v) $(-1)(-1)=1$

PROOF. For (i), note that $0a=(0+0)a=0a+0a$ by the distributive law, whence $0a$ must be the additive identity 0 by elementary group theory [Proposition 2-2, part (vi)]. A similar argument shows also that $a0=0$. For (ii), using the distributive law, we compute that

$$ab + a(-b) = a(b - b) = a0 = 0$$

whence $a(-b)$ must be $-ab$, the additive inverse of ab, and similarly for $(-a)b$. Noting that for all $a \in A$, $-(-a)=a$ [Proposition 2-2, part (iv)], the other three properties are special cases of (ii).   □

Exercises

1. Give an example of a noncommutative group of 24 elements. (*Hint:* $24 = 4!$.)

2. Give an example of a group G and a nonempty subset H of G which is closed under the operation defined on G, but is *not* a subgroup of G. (*Hint:* G must be infinite for this to occur.)

3. Show that a group G is commutative if and only if the following statement holds:

$$(st)^{-1} = s^{-1}t^{-1} \quad \forall s,t \in G$$

[*Hint*: Using Proposition 2-2, parts (iv) and (v), compute $(s^{-1}t^{-1})^{-1}$ two ways.]

4. Show that the following Cayley table can only be completed in one way so that the elements s, t, u, and v constitute a group. Deduce the requisite products, rigorously defending each step.

*	s	t	u	v
s			u	
t				
u			v	
v				

[*Hint*: What do the cancellation laws imply about the rows and columns of this table? The crux of the matter, which you may not appreciate until you have solved the problem, is that for any x in a group G, the left and right multiplication maps $y \mapsto xy$ and $y \mapsto yx$ are bijective functions from G to itself (i.e., they are permutations of the set G).]

5. Let G_0 and G_1 be groups. Consider the set

$$G_0 \times G_1 = \{(s_0,s_1) : s_0 \in G_0, s_1 \in G_1\}$$

This is just the Cartesian product of the two sets G_0 and G_1. Define an operation on $G_0 \times G_1$ as follows:

$$(s_0,s_1)(t_0,t_1) = (s_0 t_0, s_1 t_1) \quad \forall s_0,t_0 \in G_0, \ s_1,t_1 \in G_1$$

That is, we carry out the product componentwise, making use of the operations defined on the factor groups. Show that $G_0 \times G_1$ is a group with respect to this operation. Be sure to verify all requisite properties explicitly. This is called the *direct product* of G_0 and G_1.

6. Show that $G_0 \times G_1 \cong G_1 \times G_0$. (Explicitly construct the isomorphism; this is easy.)

7. Continuing in the same context, consider the functions

$$\rho_0 : G_0 \times G_1 \rightarrow G_0$$
$$(s_0, s_1) \mapsto s_0$$

$$\rho_1 : G_0 \times G_1 \rightarrow G_1$$
$$(s_0, s_1) \mapsto s_1$$

These are called *projection maps*. The first, ρ_0, retains the first coordinate and drops the second. The second, ρ_1, retains the second and drops the first. Show that both maps are surjective homomorphisms and compute the kernel of each.

8. Consider the special case of the direct product $G \times G$ of a group G with itself. Define a subset D of $G \times G$ by

$$D = \{(s,s) : s \in G\}$$

That is, D consists of all elements with both coordinates equal. Show that D is a subgroup of $G \times G$. This is called the *diagonal subgroup*. Do you see why?

9. Consider the direct product $\mathbf{R} \times \mathbf{R}$ of the additive group of real numbers with itself and the function $f: \mathbf{R} \times \mathbf{R} \rightarrow \mathbf{R}$ defined by $f(x,y) = 2x - y$. Show that f is a homomorphism of groups; describe its kernel and image.

10. Show that the function $f: \mathbf{R} \rightarrow \mathbf{R}$ defined by $f(x) = ax + b$ $(a, b \in \mathbf{R}, b \neq 0)$ is *not* a homomorphism of additive groups from $\langle \mathbf{R}, + \rangle$ to itself.

11. Let G be a group and consider the additive group of integers $\mathbf{Z}$. For any fixed $s \in G$, show that the function

$$\varphi : \mathbf{Z} \rightarrow G$$
$$n \mapsto s^n$$

is a homomorphism. Deduce from this that if G is finite, then $\mathrm{Ker}(\varphi)$ is nontrivial and therefore there exists a positive integer m such that $s^m = e$. (*Hint*: Can a map from an infinite set to a finite set be injective? Consider the answer to this question in light of Corollary 2-8.)

12. Use the previous problem and Proposition 2-6 to show that for each element s of a group G, the subset $\langle s \rangle = \{s^n : n \in \mathbf{Z}\}$ is a subgroup of G. This is called the *cyclic subgroup generated by s*.

13. Show that if H is a subgroup of G and $s \in H$, then $\langle s \rangle \subseteq H$. (*Hint*: Consider each of the three possibilities for s^n, according to whether n is positive, negative, or 0.)

14. Show that if a subgroup H of $\langle \mathbf{Z}_4, + \rangle$, the additive group of integers modulo 4, contains either 1 or 3, then in fact $H = \mathbf{Z}_4$. (See Example 8 of Section 2.1 for a refresher on $\mathbf{Z}_n$.)

15. Find all subgroups of $\langle \mathbf{Z}_5, + \rangle$, the additive group of integers modulo 5. (*Hint*: Examine each of the cyclic subgroups. Remember that a subgroup must be closed under the given operation.)

16. Let $\varphi_1, \varphi_2 : \mathbf{Z} \to G$ be homomorphisms from the additive group $\mathbf{Z}$ to an arbitrary group G. Show that if $\varphi_1(1) = \varphi_2(1)$, then $\varphi_1 = \varphi_2$. In other words, a group homomorphism from $\mathbf{Z}$ into any group is completely determined by its action on 1.

17. Prove that a permutation and its inverse have the same sign. [*Hint*: Since the sign map is a homomorphism, $\sigma(\pi^{-1}) = \sigma(\pi)^{-1}$. Now what are the multiplicative inverses of ± 1?]

18. Let A_n denote the set of all even permutations in S_n (that is, permutations with sign +1). Show that A_n is a subgroup of S_n. (*Hint*: A_n is by definition the kernel of what homomorphism?)

19. Let G be a group. For $s \in G$, define $L_s : G \to G$ by $L_s(t) = st$ for all $t \in G$. L_s is thus left multiplication (or, more properly, *left translation*) by s.

(a) Let Sym(G) denote the set of all bijections from G to itself (i.e., all permutations of G). Show that for each $s \in G$, $L_s \in$ Sym(G).

(b) Show that the mapping

$$\Lambda : G \to \mathrm{Sym}(G)$$
$$s \mapsto L_s$$

is a homomorphism of groups. This amounts to showing that $L_{st} = L_s \circ L_t$.

(c) Show that Λ is, moreover, injective. Conclude that every group is isomorphic to a subgroup of a permutation group. This result is called Cayley's Theorem. [*Hint*: To establish injectivity, compute Ker(Λ).]

20. Let A be a ring. Show that A is commutative if and only if the following identity holds:

$$(a + b)^2 = a^2 + 2ab + b^2$$

(*Hint*: Multiply out the left-hand side using the distributive law twice but not the commutative law. Compare to the right-hand side.)

21. Give an example of a commutative ring A such that neither a nor b is 0, but $ab=0$. In this case we say a and b are *zero divisors*.

22. Show that a field can never have zero divisors; that is, if $ab=0$, then either a or b is itself equal to 0.

23. Write out both Cayley tables (one for addition, and one for multiplication) for $\mathbf{F}_5$. (See Example 5 of Section 2.3 above.)

24. Write out both Cayley tables for $\mathbf{Z}_6$, the ring of integers modulo 6. Show that this ring is not a field.

25. More generally show that $\mathbf{Z}_n$ is not a field whenever n is not a prime integer. (*Hint*: Use Exercise 22 above.)

26. A commutative ring without zero divisors is called an *integral domain*. Show that a commutative ring A is an integral domain if and only if we have the cancellation law

$$ab=ac \Rightarrow b=c \quad \forall a,b,c \in A, \; a \neq 0$$

Use this result and the Pigeonhole Principle to show that every finite integral domain is a field.

27. Construct a field of four elements. (*Hint*: This might be hard. Recall that the ring $\mathbf{Z}_4$ will not work.)

3
Vector Spaces and Linear Transformations

Building on our work with groups and group homomorphisms, we now define vector spaces and linear transformations. Again the axioms may at first look arbitrary, but as we shall see in subsequent chapters, they are a masterpiece of abstraction—general enough to admit a vast range of diverse particular instances, but restrictive enough to capture the fundamental geometric notion of dimension.

3.1 Vector Spaces and Subspaces

The following is one of the two principal definitions in linear algebra. While we work over an abstract field k (see Section 2.3), the student may well think of k as the real numbers, with the understanding that the only properties we are using are the field axioms.

DEFINITION. A *vector space over a field k* consists of a set V together with a binary operation $+$ and an external scalar operation $k \times V \to V$, called *scalar multiplication*, which satisfy the following axioms:

(i) $\langle V, + \rangle$ is an additive group.

(ii) $(\lambda\mu)v = \lambda(\mu v)$ for all $\lambda, \mu \in k$, $v \in V$.

(iii) $(\lambda+\mu)v = \lambda v + \mu v$ for all $\lambda, \mu \in k$, $v \in V$.

(iv) $\lambda(v+w) = \lambda v + \lambda w$ for all $\lambda \in k$, $v, w \in V$.

(v) $1v = v$ for all $v \in V$.

In this context, elements of V are called *vectors*. Elements of k are called *scalars*, and scalar multiplication is indicated by simple juxtaposition of a scalar (on the left) with a vector. For us, the most important fields will be **R** and **C**, and we then speak, respectively, of *real* and *complex vector spaces*. Since V and k both have an additive identity, we shall distinguish them by writing **0** (bold typeface) for the zero in V and 0 (regular typeface) for the zero in k.

EXAMPLES

(1) Any field k is a vector space over itself. In this case the vector space axioms are the ordinary ring axioms applied to a field. In particular, **R** is itself a real vector space; **C** is itself a complex vector space.

(2) Let k be any field and let k^2 denote the set of all ordered pairs (x_1,x_2) of elements of k. Then k^2 is a vector space over k with respect to the operations

$$(x_1,x_2) + (y_1,y_2) = (x_1+y_1,x_2+y_2) \ \forall x_1,x_2,y_1,y_2 \in k$$

$$a(x_1,x_2) = (ax_1,ax_2) \ \forall a,x_1,x_2 \in k$$

We know that k^2 is an additive group by Exercise 5 of the previous chapter, and the other axioms are inherited directly from the field axioms for k. Note in particular that the additive group identity is $(0,0)$; the additive inverse of (x_1,x_2) is $(-x_1,-x_2)$. In the special case $k=\mathbf{R}$, we speak of the vector space $\mathbf{R}^2$ (read **R**-two), and these operations then have a clear geometric interpretation. Addition is performed by completing a parallelogram; the diagonal then represents the sum (see Figure 3.1). Scalar multiplication is just the expansion or contraction of a vector by a given factor. (Scalar multiplication by a negative number a expands or contracts by $|a|$ while also reflecting through the origin.)

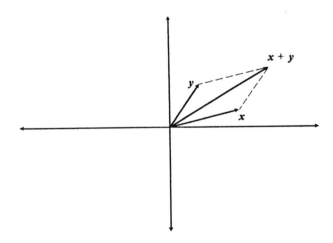

Figure 3.1. Addition in $\mathbf{R}^2$ by the parallelogram law.

(3) This example generalizes the last. Let k be any field and let n be any positive integer. Define k^n to be the set of all n-tuples of elements of k. That is,

$$k^n = \{(x_1,\ldots,x_n) : x_1,\ldots,x_n \in k\}$$

Once again define addition and scalar multiplication in k^n componentwise:

$$(x_1,\ldots,x_n) + (y_1,\ldots,y_n) = (x_1+y_1,\ldots,x_n+y_n)$$
$$a(x_1,\ldots,x_n) = (ax_1,\ldots,ax_n)$$

Then k^n is a vector space over k. The student should again carefully review the axioms. In the special case $k=\mathbf{R}$, we speak of *real n-space*, $\mathbf{R}^n$. ($\mathbf{R}^3$ has the geometric interpretation of ordinary three-dimensional space.) In the special case $k=\mathbf{C}$, we speak of *complex n-space*, $\mathbf{C}^n$. Generally in k^n we distinguish vectors from scalars by using bold type. Thus we might write

$$x = (x_1,\ldots,x_n)$$

In hand-written discourse, one also uses the notation $\vec{x}$ to denote a vector in real or complex n-space.

(4) $\mathscr{C}^0(\mathbf{R})$, the set of continuous real-valued functions defined on $\mathbf{R}$, is a real vector space with respect to ordinary addition of functions with scalar multiplication defined by

$$(af)(x) = a \cdot f(x)$$

To illustrate, we verify one of the distributive laws:

$$[a(f+g)](x) = a \cdot (f+g)(x) = a \cdot (f(x)+g(x)) = a \cdot f(x) + a \cdot g(x) = (af)(x) + (ag)(x)$$

which shows that $a(f+g) = af + ag$ for all real a. Similarly, $\mathscr{C}^n(\mathbf{R})$, the set of all real-valued functions on $\mathbf{R}$ having continuous nth derivative defined everywhere, is a real vector space. One can even replace $\mathbf{R}$ by $\mathbf{C}$ to obtain the complex vector space of complex-differentiable functions, but complex differentiability is an altogether more delicate matter.

(5) The set $\mathbf{Q}[x]$ of polynomial functions with rational coefficients is a vector space over $\mathbf{Q}$ with respect to ordinary addition and scalar multiplication of

polynomials. Similarly, one has the real vector space $\mathbf{R}[x]$ and the complex vector space $\mathbf{C}[x]$.

We will meet many more examples soon, but first we need some elementary properties common to all vector spaces. Note that all of the properties of additive groups remain in effect. In particular, the identity 0 and additive inverses are unique. Moreover, we have a cancellation law for addition.

3-1 PROPOSITION. *Let V be a vector space over a field k. Then the following assertions hold:*

(i) $\lambda 0 = 0 \quad \forall \lambda \in k$

(ii) $0v = 0 \quad \forall v \in V$

(iii) $(-\lambda)v = -(\lambda v) \quad \forall \lambda \in k,\ v \in V$

(iv) $\lambda v = 0 \Leftrightarrow (\lambda = 0 \text{ or } v = 0), \quad \forall \lambda \in k,\ v \in V$

PROOF. (i) By the right distributive law, $\lambda 0 = \lambda(0+0) = \lambda 0 + \lambda 0$, whence $\lambda 0$ must be the zero vector by elementary group theory [Proposition 2-2, part (vi)].

(ii) By the left distributive law, $0v = (0+0)v = 0v + 0v$, and the same argument as above shows that $0v = 0$.

(iii) Compute the sum $\lambda v + (-\lambda)v = (\lambda - \lambda)v = 0v = 0$. Hence $(-\lambda)v$ is indeed the additive inverse of λv, as claimed.

(iv) Suppose that $\lambda v = 0$, but $\lambda \neq 0$. Then multiplying both sides of $\lambda v = 0$ by λ^{-1}, we have, according to part (i) and the associative law for scalar multiplication, that $\lambda^{-1}(\lambda v) = (\lambda^{-1}\lambda)v = 1v = v = 0$, as required. $\quad\square$

Subspaces

As we did for groups, we now explore the corresponding substructure for vector spaces.

DEFINITION. A subset W of a vector space V over a field k is called a *subspace* of V if it constitutes a vector space over k in its own right with respect to the additive and scalar operations defined on V.

One way to establish that W is a subspace of V would be to show that it is an additive subgroup which is closed under scalar multiplication, since in this case the requisite axioms concerning scalar multiplication are inherited from V. But it turns out that there is a much simpler criterion.

3-2 PROPOSITION. (Subspace Criterion) *Let W be a nonempty subset of the vector space V. Then W is a subspace of V if and only if it is closed under addition and scalar multiplication.*

PROOF. Assume that W is closed under addition and scalar multiplication. To show that W is a subspace, it suffices to show that it is an additive subgroup, and for this we need only show that if v and w lie in W, then so does $v - w$ (Proposition 2-3). According to the previous result,

$$v - w = v + (-w) = v + (-1)w$$

But now consider the expression on the right: $(-1)w$ lies in W, since by assumption W is closed under scalar multiplication. Hence so does the sum, since by assumption W is moreover closed under addition. The converse is clear.   ☐

EXAMPLES

(1) Given any vector space V, the subsets $\{0\}$ and V itself are clearly subspaces of V. The former is called the *trivial subspace*.

(2) Consider the real vector space $V = \mathbf{R}^2$, viewed as the coordinate plane. Then both the x- and y-axes are subspaces of V. This is most easily seen geometrically: the sum of two vectors on either axis remains on that axis; a scalar multiple of a vector lying on an axis again remains on the axis. The same argument shows more generally that any line through the origin constitutes a subspace of V.

(3) Next consider the vector space $V = k^n$ where k is any field. Let W be the set of vectors $(x_1, \ldots, x_n)$ satisfying the equation

$$x_1 + \cdots + x_n = 0$$

Then W is a subspace of V. The verification is left as an exercise. Notice that this generalizes to the solution set of *any* linear equation of the form

$$a_1 x_1 + \cdots + a_n x_n = 0$$

in n variables. (What happens in the case that the right-hand side is not 0?)

(4) Working within $\mathscr{C}^0(\mathbf{R})$, we have a chain of subspaces

$$\mathscr{C}^0(\mathbf{R}) \supseteq \mathscr{C}^1(\mathbf{R}) \supseteq \mathscr{C}^2(\mathbf{R}) \supseteq \mathscr{C}^3(\mathbf{R}) \supseteq \cdots$$

Here $\mathscr{C}^n(\mathbf{R})$ denotes the set of all real-valued functions on $\mathbf{R}$ with continuous nth derivative defined everywhere. The required closure properties are familiar from calculus.

(5) Let $V=\mathbf{Q}[x]$, the rational vector space of all polynomials in x with rational coefficients, and let n be any nonnegative integer. Then the set W of all polynomials of degree less than or equal to n is a subspace of V. This holds because the sum of two polynomials of degree less than or equal to n is again a polynomial of degree less than or equal to n, and similarly for scalar multiplication. (By convention, the degree of the zero polynomial is considered to be $-\infty$.)

We now introduce a generic construction of subspaces.

DEFINITION. Let $v_1,\ldots,v_n$ be a family of vectors in the vector space V defined over a field k. Then an expression of the form

$$\lambda_1 v_1 + \lambda_2 v_2 + \cdots + \lambda_n v_n \quad (\lambda_1,\lambda_2,\ldots,\lambda_n \in k)$$

is called a *linear combination* of the vectors $v_1,\ldots,v_n$. The set of all such linear combinations is called the *span* of $v_1,\ldots,v_n$ and denoted $\mathrm{Span}(v_1,\ldots,v_n)$.

Note that we can extend the notion of span to an infinite family of vectors with the understanding that linear combinations drawn from such a collection may involve only finitely many vectors with nonzero coefficients.

3-3 PROPOSITION. *Let $v_1,\ldots,v_n$ be a family of vectors in the vector space V. Then $W=\mathrm{Span}(v_1,\ldots,v_n)$ is a subspace of V.*

PROOF. Let us consider two linear combinations

$$\lambda_1 v_1 + \lambda_2 v_2 + \cdots + \lambda_n v_n \quad (\lambda_1,\lambda_2,\ldots,\lambda_n \in k)$$

and

$$\mu_1 v_1 + \mu_2 v_2 + \cdots + \mu_n v_n \quad (\mu_1,\mu_2,\ldots,\mu_n \in k)$$

Their sum, by the distributive law, is

$$(\lambda_1+\mu_1)v_1 + (\lambda_2+\mu_2)v_2 + \cdots + (\lambda_n+\mu_n)v_n$$

which is also a linear combination of the given family. In the same vein, it is clear that a scalar multiple of a linear combination of the v_j is likewise a linear combination of the v_j. According to the subspace criterion given above, this suffices. $\square$

If $W=\mathrm{Span}(v_1,...,v_n)$, we say that W is *generated* or *spanned* by $v_1,...,v_n$. We also say that this family *spans* W. In the case that V itself is spanned by a finite collection of vectors, we say that V is *finitely generated* or *finite dimensional*.

EXAMPLES

(1) Let $V=\mathbf{R}^2$ be the coordinate plane. Then the span of the single vector $(1,1)$ consists of all multiples $a(1,1)=(a,a)$ and hence is the line $x_2=x_1$. Now consider the span of the two vectors $(1,1)$ and $(1,0)$. We claim that this is V itself. To verify the claim, we must show that any given point (x_1,x_2) can be expressed as a linear combination of $(1,1)$ and $(1,0)$. But this is to say that we can always solve the vector equation

$$a(1,1) + b(1,0) = (x_1,x_2)$$

for a and b. This in turn amounts to the linear system

$$a + b = x_1$$
$$a \quad\;\; = x_2$$

for which we indeed have the obvious solution $a=x_2$, $b=x_1-x_2$. The student would do well to interpret this result geometrically.

(2) Let $V=k^n$, where k is any field. Consider the family of n vectors

$$e_1 = (1,0,0,...,0)$$
$$e_2 = (0,1,0,...,0)$$
$$\vdots$$
$$e_n = (0,0,0,...,1)$$

These are called the *canonical basis vectors* for reasons to be explained subsequently. Clearly these vectors span V, since it is trivial to write any vector $(x_1,...,x_n)$ as a linear combination of the e_j:

$$(x_1,...,x_n) = \sum_{j=1}^{n} x_j e_j$$

It follows that V is finite dimensional.

(3) Consider the span of the functions $\sin(x)$ and $\cos(x)$ in the real vector space $\mathscr{E}^0(\mathbf{R})$. This consists of all functions of the form

$$f(x) = a\sin(x) + b\cos(x)$$

where a and b are real numbers. We shall see later, by the way, that $\mathscr{E}^0(\mathbf{R})$ is not finite dimensional.

(4) Let V be the vector space of polynomials with rational coefficients of degree less than or equal to n. (Recall that this is a subspace of $\mathbf{Q}[x]$.) Then V is finitely generated and spanned by the polynomials $1, x, x^2, \ldots, x^n$.

(5) The reader may know that the general solution to the ordinary differential equation

$$y'' - 4y = 0$$

is $y = c_0 e^{2x} + c_1 e^{-2x}$. In the language of the current discussion, this is to say that the general solution is precisely the span of the functions e^{2x} and e^{-2x} in $\mathscr{E}^2(\mathbf{R})$. In particular, the solution set constitutes a subspace of this vector space. For reasons that will become clear in the following section, such behavior is typical of homogeneous linear differential equations.

3.2 Linear Transformations

This is the second of the two principal definitions in linear algebra. It expresses a relationship between vector spaces (over the same field of scalars) which preserves elements of structure.

DEFINITION. Let V and V' be vector spaces over a common field k. Then a function $T: V \rightarrow V'$ is called a *linear transformation* if it satisfies the following conditions:

(i) $T(v+w) = T(v) + T(w)$ $\forall v, w \in V$

(ii) $T(\lambda v) = \lambda T(v)$ $\forall v \in V, \lambda \in k$

One also says that T is *k-linear* or a *vector space homomorphism*.

Note that the first condition states that T is a homomorphism of additive groups, and therefore all of our previous theory of group homomorphisms applies. In particular, we have the following derived properties:

(iii) $T(0) = 0$

(iv) $T(-v) = -T(v) \quad \forall v \in V$

(v) $T(mv) = m \cdot T(v) \quad \forall v \in V, m \in \mathbf{Z}$

EXAMPLES

(1) The constant map from V to V' that sends every element of V to the zero vector in V' is, of course, linear. This is called the *zero map*.

(2) Let V be any vector space. Then $1_V: V \to V$ is a linear transformation. More generally, let λ be any element of k. Then the function $T_\lambda: V \to V$ defined by $T_\lambda(v) = \lambda v$ is a linear transformation. This follows from the left distributive law for vector spaces and the commutativity of k.

(3) Let $V = k^n$. Define a family of maps $\rho_1, \dots, \rho_n: V \to k$ by

$$\rho_j(x_1, \dots, x_n) = x_j \quad (j=1, \dots, n)$$

Then ρ_j is called *projection onto the jth coordinate*. The student should verify that these maps are indeed linear.

(4) Let $V = \mathbf{R}^2$ and let a,b,c,d be any real numbers. Then the map $T: V \to V$ defined by

$$T(x_1, x_2) = (ax_1 + bx_2, cx_1 + dx_2)$$

is a linear transformation. To check this we first verify additivity:

$$T((x_1, x_2) + (y_1, y_2)) = T(x_1 + y_1, x_2 + y_2)$$
$$= (a(x_1 + y_1) + b(x_2 + y_2), c(x_1 + y_1) + d(x_2 + y_2))$$

$$T(x_1, x_2) + T(y_1, y_2) = (ax_1 + bx_2, cx_1 + dx_2) + (ay_1 + by_2, cy_1 + dy_2)$$

Comparing coordinates, we see that both expressions are equal. We next verify that scalar multiplication commutes with T. Let λ be any real number. Then

$$T(\lambda(x_1, x_2)) = T(\lambda x_1, \lambda x_2)$$
$$= (a\lambda x_1 + b\lambda x_2, c\lambda x_1 + d\lambda x_2)$$
$$= \lambda(ax_1 + bx_2, cx_1 + dx_2)$$
$$= \lambda T(x_1, x_2)$$

This example is most important because it shows that solving the linear system

$$ax_1 + bx_2 = y_1$$
$$cx_1 + dx_2 = y_2$$

amounts to solving the vector equation $T(x_1,x_2)=(y_1,y_2)$. Thus solving a linear system is equivalent to finding an inverse image under a linear map! This observation generalizes to all dimensions.

(5) Let $D: \mathscr{C}^1(\mathbf{R}) \to \mathscr{C}^0(\mathbf{R})$ be the differentiation operator. Then D is a linear transformation of real vector spaces. This amounts to the familiar rules of differentiation

$$D(f+g) = D(f) + D(g)$$
$$D(af) = a \cdot D(f)$$

for all $f, g \in \mathscr{C}^1(\mathbf{R})$ and $a \in \mathbf{R}$.

We resume the main exposition to consider general properties of linear transformations.

3-4 PROPOSITION. *The composition of linear transformations is a linear transformation.*

PROOF. Let $T: V \to V'$ and $T': V' \to V''$ be linear transformations. We know that their composition $T' \circ T$ is a group homomorphism, so we need only show that it commutes with scalar multiplication. This is trivial:

$$
\begin{aligned}
T' \circ T(\lambda v) &= T'(T(\lambda v)) \\
&= T'(\lambda(T(v)) \\
&= \lambda T'(T(v)) \\
&= \lambda(T' \circ T)(v)
\end{aligned}
$$
$\square$

Note that since a linear transformation is a special case of a group homomorphism, its kernel and image are defined as for groups, and these are subgroups of their respective ambient groups by Proposition 2-6. But even more is true.

3-5 PROPOSITION. *The kernel and image of a linear transformation are subspaces of their ambient vector spaces.*

PROOF. Let $T:V \rightarrow V'$ be a linear transformation. We already know that Ker(T) and Im(T) are subgroups, so it suffices to show that they are closed under scalar multiplication. Assume that v lies in Ker(T). Then for any $\lambda \in k$,

$$T(\lambda v) = \lambda T(v) = \lambda 0 = 0$$

and therefore λv also lies in Ker(T). Closure of the image under scalar multiplication is equally straightforward and left as an exercise. □

The next definition, proposition, and corollary mimic the corresponding results for groups developed in Section 2.2.

DEFINITION. A bijective linear transformation $T:V \rightarrow V'$ is called an *isomorphism of vector spaces*.

We often write $V \cong V'$ to indicate the existence of such an isomorphism. The student should show that if T is an isomorphism of vector spaces, so is the inverse map T^{-1}.

Since linear transformations are in particular group homomorphisms, the following assertions need no further proof. (See Proposition 2-7.)

3-6 PROPOSITION. *Let $T:V \rightarrow V'$ be a linear transformation and suppose that $T(v) = v'$. Then the inverse image of v' under T consists of sums of the form $v+u$ where u lies in the kernel of T. That is,*

$$T^{-1}(v') = \{v + u : u \in \text{Ker}(T)\}$$

□

3-7 COROLLARY. *A linear transformation is injective if and only if it has zero kernel.* □

We now look at some particular cases of this elementary theory.

EXAMPLES

(1) Consider the map $T:\mathbf{R}^2 \rightarrow \mathbf{R}$ defined by $T(x_1,x_2) = x_1 + x_2$. The kernel of T consists of all points of the form $(x,-x)$ and is thus a line through the origin. It follows from the corollary that T is not injective. What is the inverse image under T of any real number a? According to the previous proposition, since $(a,0)$ is one such pre-image, the entire inverse image consists of points of the form $(a,0) + (x,-x)$. This is just the line through $(a,0)$ parallel to the line defined by the kernel.

(2) We show on general principles that if the real linear system of equations

$$2x_1 - x_2 = y_1$$
$$x_1 + x_2 = y_2$$

has any solution for given y_1 and y_2, then it is unique. Recall that we can realize this system as the vector equation $T(x_1,x_2) = (y_1,y_2)$ where T is the linear transformation defined by

$$T(x_1,x_2) = (2x_1 - x_2, x_1 + x_2)$$

To show uniqueness of solutions amounts to showing that T is injective, which in turn is equivalent to showing that T has zero kernel. Hence the entire problem reduces to establishing that the associated *homogeneous* system

$$2x_1 - x_2 = 0$$
$$x_1 + x_2 = 0$$

has only the solution $(0,0)$. But this is obvious, as one sees from adding both equations to show that both x_1 and hence x_2 are 0. This type of argument admits vast generalization, as we shall see later when we treat the general topic of linear systems.

(3) Let $V = \mathbf{R}^{n+1}$ and let V' be the subspace of $\mathbf{R}[x]$ consisting of all polynomials of degree less than or equal to n. Define a function $T: V \to V'$ by

$$T(a_0,\dots,a_n) = a_0 + a_1 x^1 + \cdots + a_n x^n$$

Hence T takes the $n + 1$ coordinates of a vector in V as coefficients of a polynomial in V'. This map is evidently linear, according to the arithmetic of ordinary polynomials, and certainly invertible (map a polynomial to the vector defined by its coefficients). Hence $V \cong V'$.

(4) Let $V = \mathscr{C}^2(\mathbf{R})$ and $V' = \mathscr{C}^0(\mathbf{R})$. Then the map $T: V \to V'$ defined by

$$T(y) = y'' + y$$

is linear. (Verify!) Solving the ordinary differential equation

$$y'' + y = x^4 + 12x^2$$

amounts to finding the inverse image of $f(x)=x^4+12x^2$ under T. An obvious particular solution is $y = x^4$. Now the general solution to the corresponding homogeneous equation

$$y'' + y = 0$$

is well known to be

$$y = c_1\cos(x) + c_2\sin(x) \quad (c_1,c_2 \in \mathbf{R})$$

and this amounts to the kernel of T. Hence according to the previous proposition, the general solution to the original nonhomogeneous equation is

$$y = x^4 + c_1\cos(x) + c_2\sin(x) \quad (c_1,c_2 \in \mathbf{R})$$

This discussion is typical of linear differential equations.

3.3 Direct Products and Internal Direct Sums

We now introduce two related constructions. The first describes how two or more vector spaces over the same field may be combined into a single space. The second describes how a given vector space may be decomposed into a kind of summation of subspace components.

Although both of these notions are mathematically fundamental, this material is not used heavily in the sequel and may be omitted on a first reading.

Direct Products

Let W_0 and W_1 be vector spaces over a common field k. We define their *direct product*, $V = W_0 \times W_1$, as follows. As a set, V consists of all ordered pairs of elements from W_0 and W_1, respectively; that is,

$$V = \{(w_0, w_1) : w_0 \in W_0 \text{ and } w_1 \in W_1\}$$

The additive structure on V is given componentwise:

$$(w_0, w_1) + (u_0, u_1) = (w_0 + u_0, w_1 + u_1) \quad \forall w_0, u_0 \in W_0, w_1, u_1 \in W_1$$

And so, too, for the scalar multiplication:

$$\lambda(w_0, w_1) = (\lambda w_0, \lambda w_1) \quad \forall \lambda \in k, w_0 \in W_0, w_1 \in W_1$$

As a special case of Exercise 5 of Chapter 2, we know that $\langle V, + \rangle$ is an additive group with identity $(0,0)$. The vector space axioms for scalar multiplication are easily verified.

Clearly we can extend the notion of a direct product to more than two vector spaces. Note that the familiar vector space k^n is in fact a particular instance of this construction:

$$k^n = \underbrace{k \times k \times \ldots \times k}_{n\text{-times}}$$

Continuing with the case $V = W_0 \times W_1$, we have two *projection maps* ρ_0 and ρ_1 given by

$$\begin{array}{ccc} & \rho_0 & \\ V & \to & W_0 \\ (w_0, w_1) & \mapsto & w_0 \end{array} \qquad \begin{array}{ccc} & \rho_1 & \\ V & \to & W_1 \\ (w_0, w_1) & \mapsto & w_1 \end{array}$$

The student should verify that each of these maps is a surjective linear transformation. Again, these notions extend easily to the direct product of more than two vector spaces.

The direct product has an abstract universal property. Given a pair of linear transformations $T_0 : U \to W_0$ and $T_1 : U \to W_1$ defined on a common vector space U over k, there exists a unique linear transformation

$$T_0 \times T_1 : U \to W_0 \times W_1$$

such that

$$\rho_0 \circ (T_0 \times T_1) = T_0 \quad \text{and} \quad \rho_1 \circ (T_0 \times T_1) = T_1$$

This may be expressed by the commutative diagram shown in Figure 3.2.

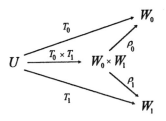

Figure 3.2. The universal property of the direct product.

The definition of $T_0 \times T_1$ (and also its uniqueness) is clear: since the jth component of $T_0 \times T_1(u)$ must be $T_j(u)$, $(j=0,1)$ we must have

$$T_0 \times T_1(u) = (T_0(u), T_1(u))$$

for all $u \in U$. We leave the routine details of linearity as an exercise and pass on to the next construction, the internal direct sum.

Internal Direct Sums

Let V be a vector space over a field k. Our task now is to identify conditions under which V can be decomposed into the sum of two or more subspaces which are completely independent of each other in a sense to be made clear presently (and even more so in the following chapter).

Let W_0 and W_1 be subspaces of V. Then we write $V = W_0 + W_1$ if every element of V can be written as a sum of the form $w_0 + w_1$ with $w_j \in W_j$ $(j=0,1)$. If, moreover, W_0 and W_1 have only the zero vector in common (i.e., $W_0 \cap W_1 = \{0\}$), we then write $V = W_0 \oplus W_1$ and say that V is the *internal direct sum* of W_0 and W_1. Internal direct sums satisfy the following key proposition, which in turn becomes the basis of a universal property strikingly related to that of the direct product.

3-8 PROPOSITION. *Let V be a vector space. Then the following three statements are equivalent:*

(i) *V is the internal direct sum of subspaces W_0 and W_1.*

(ii) *Every element $v \in V$ can be expressed uniquely in the form*

$$v = w_0 + w_1$$

with $w_j \in W_j$ $(j=0,1)$.

(iii) *$V = W_0 + W_1$, and whenever $w_0 + w_1 = 0$ for some $w_j \in W_j$ $(j=0,1)$, then both w_0 and w_1 are 0.*

PROOF. (i) $\Rightarrow$ (ii) Assume that $V = W_0 \oplus W_1$. We must show that the representation of every element in V as the sum of elements from W_0 and W_1, respectively, is unique. Suppose that

$$w_0 + w_1 = u_0 + u_1$$

where $w_j, u_j \in W_j$ $(j=0,1)$. Then

$$w_0 - u_0 = u_1 - w_1$$

But the left-hand side of this equation clearly lies in the subspace W_0, while the right-hand side lies in W_1. Hence their common value lies in $W_0 \cap W_1$ which is by assumption $\{0\}$. Thus

$$w_0 - u_0 = u_1 - w_1 = 0$$

whence $w_j = u_j$ $(j=0,1)$, as required.

(ii) $\Rightarrow$ (iii) We are given that $V = W_0 + W_1$ and that, in particular, the representation of 0 as a sum of elements in W_0 and W_1 is unique. Hence $0 = 0 + 0$ is indeed the only such decomposition.

(iii) $\Rightarrow$ (i) Assuming (iii), let $w \in W_0 \cap W_1$. Then the equation

$$0 = w + (-w)$$

implies that $w = 0$ (since trivially the first term on the right lies in W_0 and the second term lies in W_1). Hence $W_0 \cap W_1 = \{0\}$ and $V = W_0 \oplus W_1$, as claimed.   □

Just as with direct products, the notion of an internal direct sum extends easily to the case of more than two subspaces. In fact, we say V is the internal direct sum of subspaces $W_0, \ldots, W_m$ $(m \geq 1)$ and write

$$V = W_0 \oplus \cdots \oplus W_m$$

if the following conditions are satisfied:

(i) $V = W_0 + \cdots + W_m$; i.e., every element in V can be expressed as a sum of elements in the subspaces $W_0, \ldots, W_m$.

(ii) For all $j = 0, \ldots, m$, $W_j \cap \sum_{i \neq j} W_i = \{0\}$.

That is, each W_j has trivial intersection with the span of the other direct summands. With this extension of the definition, Proposition 3-8 admits an obvious and direct generalization, which we leave as an exercise.

We conclude with an abstract universal property exhibited by the direct sum. Assume that $V = W_0 \oplus W_1$. We have two inclusion maps

$$
\begin{array}{ccc}
W_0 & \overset{i_0}{\to} & V \\
w_0 & \mapsto & w_0
\end{array}
\qquad\qquad
\begin{array}{ccc}
W_1 & \overset{i_1}{\to} & V \\
w_1 & \mapsto & w_1
\end{array}
$$

which are obviously linear transformations. Now let there be given linear transformations $T_0 : W_0 \to U$ and $T_1 : W_1 \to U$ into a common vector space U over k.

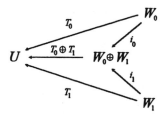

Figure 3.3. The universal property of the direct sum.

Then there exists a unique linear transformation

$$T_0 \oplus T_1 : W_0 \oplus W_1 \to U$$

such that

$$(T_0 \oplus T_1) \circ i_0 = T_0 \quad \text{and} \quad (T_0 \oplus T_1) \circ i_1 = T_1$$

This may be expressed by the commutative diagram shown in Figure 3.3.

Note how this diagram relates to the corresponding picture for the direct product: the arrows have been reversed! For this reason one says the direct sum and the direct product are *dual* constructions.

The definition of $T_0 \oplus T_1$ (and hence its uniqueness) is straightforward. Given any $v \in V$, we may write v uniquely as a sum $w_0 + w_1$ with $w_j \in W_j$ ($j=0,1$). Now set

$$T_0 \oplus T_1(w_0 + w_1) = T_0(w_0) + T_1(w_1)$$

For $w_0 \in W_0$, $i_0(w_0) = w_0 + 0$ (expressed as a sum of vectors from W_0 and W_1), so

$$(T_0 \oplus T_1) \circ i_0(w_0) = T_0(w_0) + T_1(0) = T_0(w_0)$$

Similarly,

$$(T_0 \oplus T_1) \circ i_1(w_1) = T_1(w_1)$$

as required. Again we leave the routine details of linearity as an exercise.

EXAMPLES

(1) Here is a simple example of a direct sum, upon which we shall expand implicitly in the following chapter. Let $V = \mathbf{R}^2$ and define subspaces W_0 and W_1 of V as follows:

$$W_0 = \{(x,0) : x \in \mathbf{R}\}$$

$$W_1 = \{(0,y) : y \in \mathbf{R}\}$$

Then W_0 is precisely the x-axis, while W_1 is precisely the y-axis. We have the obvious facts

(i) $V = W_0 + W_1$;

(ii) W_0 and W_1 intersect only at 0.

Therefore $V = W_0 \oplus W_1$.

(2) Let $V = \mathbf{R}[x]$, the space of all real polynomials in x. Define two subspaces of V as follows:

$$W_0 = \text{Span}\{1, x^2, x^4, \ldots\}$$
$$W_1 = \text{Span}\{x, x^3, x^5, \ldots\}$$

Recall that elements of W_0 are called *even* polynomials, and elements of W_1 are called *odd* polynomials. Since every polynomial in V can be expressed uniquely as the sum of an even polynomial and an odd polynomial, it follows that $V = W_0 \oplus W_1$.

Exercises

1. Show that the solution set W of vectors (x_1, x_2) in $\mathbf{R}^2$ satisfying the equation

    $$x_1 + 8x_2 = 0$$

    is a subspace of $\mathbf{R}^2$.

2. Determine whether the solution set to the equation

    $$x_1^2 + x_2^2 + x_3^2 = 1$$

    is a subspace of $\mathbf{R}^3$. (*Hint*: What element must a subspace contain?)

3. Let $V = \mathbf{R}^2$. Exhibit a subset of V which is an additive subgroup of V, but *not* a subspace of V.

4. Show that a subspace of $\mathbf{R}^2$ containing both $(1,0)$ and $(1,-1)$ must be all of $\mathbf{R}^2$ itself.

5. Let V be a vector space over a field k and suppose that U and W are subspaces of V. Show that $U \cap W$, the intersection of U and W, is likewise a subspace of V.

6. Show by example, using $V=\mathbf{R}^2$ for instance, that if U and W are subspaces of V, then $U \cup W$, the union of U and W, is not necessarily a subspace of V.

7. Continuing in the context of the previous problem, show that $U \cup W$ is a subspace of V if and only if either $U \subseteq W$ or $W \subseteq U$.

8. Again let U and W be subspaces of V. Let $U+W$ denote the set of all vectors of the form $u+w$ where $u \in U$ and $w \in W$. Show that $U+W$ is a subspace of V.

9. Continuing in the context of the previous problem, show that $U+W$ is the smallest subspace of V containing both U and W; that is, if X is any subspace of V containing both U and W, then $U+W \subseteq X$.

10. Let $V=\mathbf{R}^2$. What is the span of the vectors $(2,4)$ and $(-5,-10)$? Describe this set geometrically.

11. Let $V=\mathbf{R}^2$. What is the span of the vectors $(1,2)$ and $(2,1)$? Describe this set geometrically.

12. Let $V=\mathbf{R}^2$. State geometrically under what conditions the span of two non-zero vectors x and y is all of V.

13. Is it possible to find a subspace of $\mathbf{R}^2$ which is neither a point, nor a line, nor all of $\mathbf{R}^2$ itself? Explain geometrically.

14. Show that the functions e^x, $\cos(x)$, and $\sin(x)$ in $\mathscr{C}^0(\mathbf{R})$ are not in the span of the infinite family of monomials $1, x, x^2, \ldots$. Remember that a linear combination, even when drawn from an infinite family, may nonetheless involve only finitely many terms. (*Hint*: This problem essentially asks you to show that these familiar transcendental functions cannot be expressed as polynomials. What is the fate of a polynomial of degree n after being differentiated $n+1$ times?)

15. Show that at least two vectors are required to span the vector space $\mathbf{R}^2$.

16. Let $a \in \mathbf{R}$ and consider the function

$$f:\mathbf{R} \to \mathbf{R}$$

$$x \mapsto ax$$

Show that f is a linear transformation of real vector spaces. What is the kernel of f? The image? (Be sure to distinguish two cases according to whether $a=0$.)

17. Define a function A from $\mathbf{R}^n$ to $\mathbf{R}$ as follows:

$$A(x) = \frac{1}{n}\sum_{j=1}^{n}x_j$$

Thus $A(x)$ is just the average of the components of x. Show that A is a linear transformation.

18. Express the kernel of the linear transformation $T : \mathbf{R}^2 \to \mathbf{R}$ defined by $T(x_1,x_2) = 2x_1 - 5x_2$ as the span of a single vector.

19. What is the kernel of the second derivative operator

$$\mathscr{C}^2(\mathbf{R}) \to \mathscr{C}^0(\mathbf{R})$$

$$f \mapsto \frac{d^2 f}{dx^2}$$

on the real vector space of twice differentiable functions with continuous second derivative?

20. What is the kernel of the nth derivative operator

$$\mathscr{C}^n(\mathbf{R}) \to \mathscr{C}^0(\mathbf{R})$$

$$f \mapsto \frac{d^n f}{dx^n}$$

on the real vector space of n-times differentiable functions with continuous nth derivative?

21. Explain succinctly why the solution space of the differential equation

$$y'' - 2y' + y = 0$$

is a subspace of $\mathscr{C}^2(\mathbf{R})$.

22. Show that a linear map $T : \mathbf{R} \to \mathbf{R}^2$ cannot be surjective.

23. Let $a \in \mathbf{R}$ be fixed and consider the function v_a defined by

$$\mathscr{C}^0(\mathbf{R}) \overset{v_a}{\to} \mathbf{R}$$

$$f \mapsto f(a)$$

This map is called *evaluation at a*. Note how it reverses the usual roles of function and variable. Show that for all real numbers a, the evaluation map v_a is a linear transformation and describe $\mathrm{Ker}(v_a)$. (For the novice, this important construction can resemble an Escher drawing!)

24. Let a and b be real numbers. Define a function $I : \mathscr{C}^0(\mathbf{R}) \to \mathbf{R}$ on the space of continuous real-valued functions on $\mathbf{R}$ as follows:

$$I(f) = \int_a^b f(x)\,dx$$

Show that I is a linear transformation. Deduce from this that the set

$$\left\{ f \in \mathscr{C}^0(\mathbf{R}) : \int_a^b f(x)\,dx = 0 \right\}$$

is a subspace of $\mathscr{C}^0(\mathbf{R})$. Assuming that $a \neq b$, what is the image of I?

25. Let $T : V \to V$ be a linear transformation of vector spaces over k and let $\lambda \in k$ be a scalar such that there exists a nonzero $v \in V$ satisfying the equation

$$T(v) = \lambda v$$

Thus T maps v onto a scalar multiple of itself. (Note that we do not claim this identity for all elements of V, just for some of them.) Then λ is called an *eigenvalue of T* and any v (even $\mathbf{0}$) satisfying the equation above is called an *eigenvector belonging to λ*. Show that the set of all eigenvectors belonging to the eigenvalue λ is a subspace of V. This is called the *eigenspace* belonging to λ.

26. Given a vector space V over k, assume that there exist subspaces W_0 and W_1 such that $V = W_0 + W_1$ (see the discussion of internal direct sums in Section 3.3). Show that if both W_0 and W_1 are finitely generated, then V is likewise finitely generated.

27. Define subspaces W_0 and W_1 in $\mathbf{R}^2$ as follows:

$$W_0 = \mathrm{Span}\ \{(0,1)\} \quad \text{and} \quad W_1 = \mathrm{Span}\ \{(1,1)\}$$

Show that $\mathbf{R}^2 = W_0 \oplus W_1$. (*Hint*: Using Proposition 3-8, this reduces to showing that a particular system of two linear equations in two unknowns has a unique solution.)

28. Let U, W_0, and W_1 be vector spaces over k and let there be given linear transformations $T_0 : U \rightarrow W_0$ and $T_1 : U \rightarrow W_1$. According to Section 3.3, we have a linear transformation $T_0 \times T_1 : U \rightarrow W_0 \times W_1$ defined by

$$T_0 \times T_1 (u) = (T_0(u), T_1(u))$$

Prove that $\text{Ker}(T_0 \times T_1) = \text{Ker}(T_0) \cap \text{Ker}(T_1)$.

29. Let $T : V \rightarrow V$ be a linear transformation from a vector space V to itself such that $T \circ T = T$. (Such a transformation is called *idempotent*.) Show that

$$V = \text{Ker}(T) \oplus \text{Im}(T)$$

[*Hint*: Given $v \in V$, consider the effect of T on $v - T(v)$.]

4
Dimension

This chapter covers the fundamental structure theory for vector spaces. In particular, we shall show that all vector spaces admit coordinate systems (called *bases*) and that the number of coordinates is intrinsic to the space. These results in turn allow us to define the *dimension* of a vector space and thus to recast this most basic of geometric notions in purely algebraic terms. In so doing, we extend the application of this concept to many settings that have no obvious *a priori* geometric interpretation.

4.1 Bases and Dimension

This first definition is somewhat subtle, but the patient reader will see shortly that it is one of the keys to defining a coordinate system for an abstract vector space.

DEFINITION. Let V be a vector space over k. Then a family of vectors $v_1,\ldots,v_n$ is called *linearly independent* if

$$\lambda_1 v_1 + \cdots + \lambda_n v_n = 0 \Rightarrow \lambda_1,\ldots,\lambda_n = 0 \quad (\lambda_j \in k, j = 1,\ldots,n)$$

Otherwise we say that the family is *linearly dependent*.

To paraphrase, linear independence asserts that the only linear combination of the v_j that yields 0 is the one for which all coefficients are 0. Linear dependence asserts that there exist coefficients $\lambda_1,\ldots,\lambda_n$, not all 0, such that

$$\lambda_1 v_1 + \cdots + \lambda_n v_n = 0$$

Note that the zero vector is never part of a linearly independent family. In the obvious sense we may also speak of a linearly independent or dependent set. (Technically, a family is always indexed and hence may have repeated elements; but see Exercise 1.) The null set is then vacuously linearly independent.

For simplicity we have given the definitions above for a finite collection of vectors, but they are equally sensible for infinite families, provided that we maintain the convention that a linear combination drawn from an infinite family may only involve finitely many terms with nonzero coefficients.

EXAMPLES

(1) The following vectors in $\mathbf{R}^3$ are linearly independent:

$$(1,0,0), (1,1,0), (1,1,1)$$

To see this, assume that we have a linear combination that sums to 0; this is to say that we have a vector equation of the form

$$a_1(1,0,0) + a_2(1,1,0) + a_3(1,1,1) = (0,0,0)$$

Then comparing components, we see first that a_3 must be 0, second that a_2 must be 0, and finally that a_1 must be 0.

(2) The following vectors in $\mathbf{R}^3$ are linearly dependent:

$$(1,0,0), (1,1,0), (2,1,0)$$

This is clear since we have the *linear dependence relation*

$$1 \cdot (1,0,0) + 1 \cdot (1,1,0) + (-1) \cdot (2,1,0) = (0,0,0)$$

The point is that we have achieved 0 as a *nontrivial* linear combination of the given vectors (the *trivial* case being where all coefficients are 0).

(3) In k^n, the canonical basis vectors $e_1,\dots,e_n$ are clearly linearly independent since

$$\sum_{j=1}^{n} \lambda_j e_j = (\lambda_1,\dots,\lambda_n)$$

and this can be 0 if and only if all of the coefficients are zero.

(4) Let v and w be any two vectors in an arbitrary vector space V. We leave it as an exercise to show that they are linearly dependent if and only if one is a scalar multiple of the other. In the special case $V = \mathbf{R}^n$, this says that two vectors constitute a linearly dependent set if and only if they are collinear. What is the analogous statement for three vectors in $\mathbf{R}^n$?

Observe that if a collection of vectors is linearly dependent, then so is any larger collection that contains it. Similarly, if a collection of vectors is linearly independent, then so is any subcollection.

We now come to an essential proposition, one that characterizes the notion of linear dependence and, in part, explains the terminology.

4-1 PROPOSITION. (Characterization of Linear Dependence) *Let $v_1,...,v_n$ be a collection of vectors. Then this family is linearly dependent if and only if one of the v_j can be expressed as a linear combination of the others.*

PROOF. $\Rightarrow$) Suppose that the v's are linearly dependent. Then there exists a linear dependence relation

$$\lambda_1 v_1 + \cdots + \lambda_n v_n = 0$$

where not all of the coefficients are 0. We may assume without loss of generality that λ_1 is nonzero. (Any other index will do just as well.) Then

$$\lambda_1 v_1 = -\lambda_2 v_2 - \cdots - \lambda_n v_n$$

and therefore

$$v_1 = -\frac{\lambda_2}{\lambda_1} v_2 - \cdots - \frac{\lambda_n}{\lambda_1} v_n$$

Thus we have indeed expressed v_1 as a linear combination of the others. Note carefully where we have used the assumption that λ_1 is nonzero.

$\Leftarrow$) Assume without loss of generality that v_1 can be written as a linear combination of the remaining v_j. Then we have an equation of the form

$$v_1 = \lambda_2 v_2 + \cdots + \lambda_n v_n$$

This yields at once the dependence relation

$$v_1 - \lambda_2 v_2 - \cdots - \lambda_n v_n = 0$$

which completes the proof.   ☐

The next result is slightly technical, but absolutely essential and used repeatedly. The idea is that a linear dependence relation among a collection of vectors implies a certain redundancy *vis-à-vis* their span.

4-2 LEMMA. *Let $v_1,\ldots,v_n$ be a collection of vectors in V. Then if there is a linear dependence relation involving v_j (i.e., v_j occurs with nonzero coefficient), the span of the collection is unchanged when v_j is deleted. In particular, v_j lies in the span of the remaining vectors.*

PROOF. As in the previous proof, we can solve for v_j in terms of the other vectors. We may then substitute the resulting expression into any linear combination in which v_j occurs. This yields an equivalent linear combination of the remaining vectors, and hence the span indeed remains unchanged when v_j is deleted.   □

This brings us to the key definition in establishing a "coordinate system" for an abstract vector space.

DEFINITION. A (possibly infinite) collection of vectors B in a vector space V is called a *basis* for V if the following two conditions are satisfied:

(i) B is linearly independent;

(ii) B spans V.

As a quick example, note that the canonical basis $e_1,\ldots,e_n$ for k^n is obviously a basis in the sense just defined.

We shall see below that conditions (i) and (ii) taken together imply a precise balance: B must contain enough vectors to span V, but not so many that it becomes linearly dependent. First we show that a basis is indeed akin to a coordinate system.

4-3 PROPOSITION. *B is a basis for V if and only if every vector in V can be written as a linear combination of the vectors in B in exactly one way.*

PROOF. ⇒) Since by assumption B spans V, every vector in V can be written somehow as a linear combination of elements of B. To demonstrate uniqueness, assume that two linear combinations of elements v_j in B ($j=1,\ldots,n$) are equal:

$$\lambda_1 v_1 + \cdots + \lambda_n v_n = \mu_1 v_1 + \cdots + \mu_n v_n$$

Moving all of the terms to the left, we obtain

$$(\lambda_1 - \mu_1)v_1 + \cdots + (\lambda_n - \mu_n)v_n = 0$$

But the v_j are linearly independent. Therefore all of the coefficients on the left must be 0, showing that the corresponding λ_j and μ_j are indeed identical.

$\Leftarrow$) First note that B clearly spans V. Now if every vector in V can only be written as a linear combination of elements of B in one way, then, in particular, $\mathbf{0}$ can only be written as a linear combination of the elements of B in one way: all coefficients must be 0. Hence by definition, B is linearly independent and therefore a basis for V. $\qquad\square$

Given that V has a basis B of n elements $v_1,\dots,v_n$, this proposition gives rise to an important linear transformation, the *coordinate map* $\gamma_B : V \to k^n$, defined as follows:

$$v = \lambda_1 v_1 + \cdots + \lambda_n v_n \implies \gamma_B(v) = (\lambda_1,\dots,\lambda_n)$$

That is, the coordinate map extracts the coefficients arising in the unique expression of v as a linear combination of the basis elements and deposits them into the corresponding components of a vector in k^n. The student should verify that γ_B is linear and is in fact an isomorphism of vector spaces. (*Hint*: The inverse map is obvious.) The coordinate map will become more and more important in later chapters.

EXAMPLES

(1) As noted above, the canonical basis for k^n is a basis. Clearly every element of k^n can be expressed uniquely as a linear combination of the canonical basis vectors as follows:

$$(\lambda_1,\dots,\lambda_n) = \sum_{j=1}^{n} \lambda_j e_j$$

With respect to this basis, the coordinate map is precisely the identity map!

(2) Consider the vector space $V = \mathbf{R}^2$. One verifies easily that the vectors

$$v_1 = (1,0) \text{ and } v_2 = (1,1)$$

both span V and are linearly independent. They therefore constitute a basis B for V. Let us consider the coordinate map. What is $\gamma_B(2,1)$? We first express $(2,1)$ as a linear combination of v_1 and v_2:

$$(2,1) = 1 \cdot (1,0) + 1 \cdot (1,1)$$

Both coordinates are thus 1 and we have

$$\gamma_B(2,1) = (1,1)$$

As an easy exercise, the student should deduce a general formula for the value of $\gamma_B(x_1, x_2)$.

(3) The vector space $V = \mathbf{R}[x]$ of polynomials with real coefficients has infinite basis $1, x, x^2, \ldots, x^n, \ldots$ since a polynomial can evidently be written in exactly one way as a linear combination of these monomials. If we take W to be the subspace of V consisting of polynomials of degree less than or equal to n, then we have the finite basis

$$B = \{1, x, x^2, \ldots, x^n\}$$

and the associated coordinate map γ_B is just the vector of coefficients in the vector space k^{n+1}:

$$\gamma_B(a_0 + a_1 x^1 + \cdots + a_n x^n) = (a_0, a_1, \ldots, a_n)$$

(This is one of many cases for which it is more sensible to label the first coordinate a_0.)

We next develop other characterizations of bases. First, two definitions:

DEFINITIONS. A subset S of V is called a *maximally linearly independent set* if it is linearly independent and not properly contained in any larger linearly independent set. A subset S of V is called a *minimal generating set* or a *minimal spanning set* if it spans V, but no proper subset of S also spans V.

4-4 PROPOSITION. *Let S be a subset of the vector space V. Then the following three statements are equivalent:*

(i)   *S is a maximally linearly independent set.*

(ii) *S is a minimal generating set.*

(iii) *S is a basis for V.*

PROOF. We show (i) $\Leftrightarrow$ (iii) and (ii) $\Leftrightarrow$ (iii), which clearly establishes the result.

(i) $\Rightarrow$ (iii) We must show that S spans V. Given $v \in V$, by assumption the set consisting of v and S is linearly dependent. Hence there is a dependence relationship in $S \cup \{v\}$ and this must involve v, since S alone is an independent set. But then v lies in the span of S by Lemma 4-2, and since v was arbitrary, S spans V, as required.

(iii) $\Rightarrow$ (i) Since S is a basis, anything we might adjoin to it can already be written as a linear combination of the elements of S, and so results in a linearly dependent set. Thus S is maximally linearly independent.

(ii) $\Rightarrow$ (iii) We must show that S is linearly independent. If not, then S is linearly dependent and we can delete an element without diminishing its span. But this contradicts the assumed minimality of S.

(iii) $\Rightarrow$ (ii) We must show that no proper subset of S can also span V. But if this were false, it would be possible to delete a vector v from S and still to span V. Then, in particular, v itself would have to lie in the span of the remaining vectors, contradicting the assumed linear independence of S (by Proposition 4-1). $\square$

We now have three characterizations of bases: as coordinate systems, as maximally linearly independent sets, and as minimal generating sets. Our next pair of results does some essential counting. We begin with a major theorem on finitely-generated vector spaces.

4-5 THEOREM. (The Exchange Theorem) *Suppose that the collection* $v_1, \ldots, v_n$ *spans V and* $w_1, \ldots, w_r$ *is a linearly independent set. Then* $r \leq n$.

PROOF. This is subtle. Consider the slightly extended family

$$v_1, \ldots, v_n, w_1$$

This family is linearly dependent, since the v's already span V and so, in particular, w_1 can be written as a linear combination of them. Now any linear dependence relation must involve some of the v's since w_1 by itself is a linear independent set. Hence by Lemma 4-2, we can delete one of the v's—let's say v_1—without diminishing the span of the collection, which accordingly remains all of V. Next consider the family

$$v_2, \ldots, v_n, w_1, w_2$$

Repeating the same argument, we find (i) that the family is linearly dependent, (ii) that the dependence relation must involve one of the v's since w_1 and w_2 are by themselves linearly independent (being a subset of a linearly independent set), and (iii) that we can delete, say, v_2 from the collection, preserving the span (which is again V). Now we can continue this process until all of the w's have been fed into the hybrid family of v's and w's, and at each step we are guaranteed to be able to delete one of the v's. But for this to be possible, r, the number of w's, must be less than or equal to n, the number of v's. $\square$

This theorem shows us, at last, that the vector space axioms suffice to capture the notion of dimension.

4-6 COROLLARY. (Existence of Dimension) *Suppose that B and B' are both bases for the finitely-generated vector space V. Then B and B' have the same number of elements.*

PROOF. First note that according to the preceding theorem, both bases must be finite since V is spanned by finitely many vectors. So let n be the number of elements in B and let n' be the number of elements in B'. Then since B is linearly independent and B' spans V, we have $n \leq n'$. But likewise B' is linearly independent and B spans V, and so $n' \leq n$. Taken together, these inequalities imply that $n=n'$.   ❑

This result in turn yields a critical definition.

DEFINITION. The number of elements in a basis for a finitely-generated vector space V is called the *dimension* of V and denoted dim(V). The zero vector space is assigned dimension 0 and the empty set as basis.

EXAMPLES

(1) The vector space k^n evidently has dimension n, since the canonical basis has n elements.

(2) The vector space $\mathbf{R}[x]$ is not finitely generated (why?), and thus we say it is *infinite dimensional*. The subspace of all polynomials of degree less than or equal to n is finite dimensional and has dimension $n+1$, according to Example 3 above.

(3) We remarked earlier that the solution set to the differential equation

$$y'' + y = 0$$

is spanned by the linearly independent functions sin(x) and cos(x). Hence this subspace of $\mathscr{C}^2(\mathbf{R})$ has dimension 2.

(4) The solution set in $\mathbf{R}^2$ to the linear equation

$$x_1 + x_2 = 0$$

is spanned by the single vector (1,–1), since the general solution is clearly $\{(x,-x) : x \in \mathbf{R}\}$. Hence its dimension is 1.

4.2 Vector Spaces Are Free

We address the existence of bases in general, beginning with a technical proposition that applies to finitely generated vector spaces and is valuable in its own right.

4-7 PROPOSITION. *Let S and S' be finite subsets of V such that*

(i) $S \subseteq S'$;

(ii) *S is linearly independent;*

(iii) *S' spans V.*

Then there exists a basis B of V such that $S \subseteq B \subseteq S'$.

In other words, between every linearly independent set and spanning set there is a basis.

PROOF. Let S consist of the linearly independent vectors $v_1, \ldots, v_n$, and suppose that S' consists of these together with the additional vectors $w_1, \ldots, w_m$. If S' is linearly independent, we are done. If not, at least one of the w_j is involved in a linear dependence relation (otherwise we contradict the independence of the v_j). By Lemma 4-2, this vector may be deleted without affecting the span, which is therefore still V. Continuing in this way to delete the redundant w's, we must eventually reach a linearly independent set that spans V. This is the desired basis B. ☐

This proposition leads at once to the fundamental structure theorem for vector spaces. We prove it here only for finite-dimensional spaces, but it is true in general.

4-8 THEOREM. *The following statements hold in any vector space:*

(i) *Every linearly independent set may be extended to a basis.*

(ii) *Every spanning set may be contracted to a basis.*

(iii) *Every vector space has a basis.*

For reasons explained in Section 6.4, the last assertion is often stated thus:

Vector spaces are free!

and hence the title of this section. One cannot overstate the consequences of this fact, which we take as the Fundamental Theorem of Linear Algebra.

PROOF. As remarked above, we shall only treat the case of a vector space V spanned by finitely many elements. Hence any linearly independent set is finite and any spanning set admits a finite subset which also spans. (See Exercise 18.)

(i) Let S be any linearly independent set and let T be a finite spanning set. Apply the previous proposition to S and $S' = S \cup T$ to obtain a basis which contains S.

(ii) Let T be any spanning set. By the introductory remarks above, we may assume that T is a finite set. Apply the previous proposition to $S' = T$ and $S = \varnothing$ to obtain a basis which is a subset of T.

The final statement follows at once from either (i) or (ii): we may extend the null set to a basis or contract any spanning set to a basis. ☐

4-9 COROLLARY. *Every finite-dimensional vector space V over k is isomorphic to k^n for some n.*

PROOF. As observed above, the coordinate map γ_B for any basis B provides the isomorphism. (The point is that we now know a basis exists.) ☐

The contractibility of spanning sets to bases leads to our last characterization of bases. First we state a proposition which constitutes an important general criterion for deciding linear independence or span.

4-10 PROPOSITION. *Let V have dimension n. Then* (i) *no subset of V of more than n vectors can be linearly independent and* (ii) *no subset of V of fewer than n vectors can span V.*

PROOF. The first assertion restates the Exchange Theorem. For the second, note that any spanning set of fewer than n elements can be contracted (by Theorem 4-8) to a basis of fewer than n elements, contradicting the uniqueness of dimension. ☐

4-11 THEOREM. *Let V have dimension n and let S be a collection of n vectors in V. Then the following three statements are equivalent:*

(i)   *S is linearly independent.*

(ii) *S spans V.*

(iii) *S is a basis for V.*

PROOF. (i) $\Rightarrow$ (iii) The first part of the previous proposition implies that S is maximally linearly independent and hence a basis for V.

(ii) $\Rightarrow$ (iii) The second part of the previous proposition implies that S is a minimal generating set and hence a basis for V.

The implications (iii) $\Rightarrow$ (i) and (iii) $\Rightarrow$ (ii) are trivial, so we have (i) $\Leftrightarrow$ (iii) and (ii) $\Leftrightarrow$ (iii), which completes the proof. $\square$

REMARK. Note the following immediate consequence of this last theorem: if V has dimension n and W is a subspace of V also of dimension n, then $W=V$. This holds because a basis for W consists of n linearly independent vectors in V, which therefore constitute a basis for V.

We conclude this hoard of fundamental results with an obvious-sounding, but slightly subtle fact about subspaces of finite-dimensional spaces.

4-12 PROPOSITION. *Every subspace of a finite-dimensional vector space is finite dimensional of smaller or equal dimension.*

PROOF. Let W be a subspace of the n-dimensional vector space V. Expand the null set one vector at a time to a maximally linearly independent set in W, which is therefore also a basis for W. This process must terminate in less than or equal to n steps, since a linearly independent set in W is also a linearly independent set in V, and, as we have seen, V cannot accommodate more than n linearly independent vectors. Hence W has a basis of less than or equal to n elements and corresponding finite dimension. [Note that we could not directly appeal to the result on the existence of bases (Theorem 4-8) since we have only proven it for finitely-generated spaces, and until now we had not shown that a subspace of a finitely-generated vector space is finitely generated.] $\square$

EXAMPLES

(1) We know now on general principles that since $\mathbf{R}^2$ has dimension 2, any three vectors in $\mathbf{R}^2$ must be linearly dependent. We can visualize this geometrically: two linearly independent vectors already span the plane. More generally, for any field k, any $n+1$ vectors in k^n must likewise be linearly dependent.

(2) The vector space $\mathscr{C}^0(\mathbf{R})$ is infinite dimensional since the polynomial functions constitute an infinite-dimensional subspace. [Actually the identification of $\mathbf{R}[x]$ with a subspace of $\mathscr{C}^0(\mathbf{R})$ requires a little work. How does one know that distinct real polynomials produce distinct functions? This is not true, for instance, over the finite field $\mathbf{F}_p$ (cf. Section 2.3). We leave this question as an exercise with the following hint: If $p(x)$ is a polynomial and $p(r)=0$, then by high school algebra $(x-r)$ is a factor of $p(x)$. Hence the only polynomial which evaluates to 0 everywhere is the 0 polynomial.]

(3) The following vectors are easily shown to be linearly independent in $\mathbf{R}^3$ and hence a basis:

$$(1,0,0), (1,1,0), (2,0,1)$$

It follows that any vector in $\mathbf{R}^3$ can be expressed uniquely as a linear combination of these three vectors.

(4) Consider the following linear system in three variables over any field:

$$a_1 x_1 + b_1 x_2 + c_1 x_3 = y_1$$
$$a_2 x_1 + b_2 x_2 + c_2 x_3 = y_2$$
$$a_3 x_1 + b_3 x_2 + c_3 x_3 = y_3$$

With respect to this system, we claim that the following statements are equivalent:

(i) The system has at least one solution for all y_1, y_2, y_3.

(ii) The corresponding homogeneous equation (all $y_j = 0$) has only the *trivial* solution (all $x_j = 0$).

(iii) The system has exactly one solution for all y_1, y_2, y_3.

To see this, note that statement (i) is equivalent to the assertion that the vectors (a_1, a_2, a_3), (b_1, b_2, b_3), (c_1, c_2, c_3) span k^3. (This is easier to see if we regard these vectors as columns rather than rows—something that we shall do more and more in the sequel.) Statement (ii) is equivalent to the assertion that (a_1, a_2, a_3), (b_1, b_2, b_3), (c_1, c_2, c_3) are linearly independent. Finally, statement (iii) is equivalent to the assertion that (a_1, a_2, a_3), (b_1, b_2, b_3), (c_1, c_2, c_3) constitute a basis for $\mathbf{R}^3$. But according to Theorem 4-11, for any three vectors in k^3, the attributes of being a spanning set, a linearly independent set, or a basis are equivalent, and therefore so are statements (i), (ii), and (iii), as claimed. This argument generalizes directly to higher dimensions, as we shall see in the following chapter.

(5) Along the same lines as (4), one observes that a homogeneous linear system of three equations in four or more unknowns always has a nontrivial solution in k^3. This reduces to the observation that four or more vectors in k^3 are linearly dependent. Again we have a more general statement in k^n.

4.3 Rank and Nullity

This brief section describes a key connection between the image and the kernel of a linear transformation of finite-dimensional vector spaces.

4-13 THEOREM. *Let $T: V \to W$ be a linear transformation of finite-dimensional vector spaces. Then*

$$\dim(\mathrm{Ker}(T)) + \dim(\mathrm{Im}(T)) = \dim(V)$$

That is, the dimension of the kernel plus the dimension of the image is equal to the dimension of the domain.

The number $\dim(\mathrm{Ker}(T))$ is often called the *nullity* of T while $\dim(\mathrm{Im}(T))$ is often called the *rank* of T. Hence this theorem is often called the Rank-Nullity Theorem.

PROOF. Let $v_1, \ldots, v_n$ be a basis for $\mathrm{Ker}(T)$. Extend this to a basis

$$v_1, \ldots, v_n, w_1, \ldots, w_m$$

for V. By definition $\dim(\mathrm{Ker}(T)) = n$ and $\dim(V) = n + m$. Hence it remains to show that $\dim(\mathrm{Im}(T)) = m$. We do this by exhibiting a basis for $\mathrm{Im}(T)$ of precisely m elements.

Since T is a surjective map onto its image, we have, by Exercise 15 below, that T maps the spanning set $v_1, \ldots, v_n, w_1, \ldots, w_m$ for V onto the spanning set

$$T(v_1), \ldots, T(v_n), T(w_1), \ldots, T(w_m)$$

for $\mathrm{Im}(T)$. Now by construction, the first n of these images are 0 (because the vectors v_j lie in the kernel of T), so we may immediately contract this spanning set to $T(w_1), \ldots, T(w_m)$. We next show that this family is also linearly independent and therefore a basis for $\mathrm{Im}(T)$. Suppose that

$$\lambda_1 T(w_1) + \cdots + \lambda_m T(w_m) = 0$$

Then by linearity, we have that

$$T(\lambda_1 w_1 + \cdots + \lambda_m w_m) = 0$$

whence

$$\lambda_1 w_1 + \cdots + \lambda_m w_m \in \mathrm{Ker}(T)$$

But since $\mathrm{Ker}(T)$ is spanned by the v_j's, it follows that

$$\lambda_1 w_1 + \cdots + \lambda_m w_m = \mu_1 v_1 + \cdots + \mu_n v_n$$

for some family of μ_j's in k. But unless all of the λ_j's and μ_j's are 0, this violates the uniqueness of coordinates relative to the basis $v_1, \ldots, v_n, w_1, \ldots, w_m$ for V (Proposition 4-3). This completes the proof. $\square$

4-14 COROLLARY. *Let $T: V \to W$ be a linear transformation of finite-dimensional spaces of the same dimension. Then the following three statements are equivalent:*

(i) *T is injective.*

(ii) *T is surjective.*

(iii) *T is an isomorphism.*

PROOF. Clearly it is enough to show (i) $\Leftrightarrow$ (ii). Let n be the common dimension of V and W. By Corollary 3-7, T is injective if and only if $\mathrm{Ker}(T) = \{0\}$; which is to say, if and only if $\dim(\mathrm{Ker}(T)) = 0$. By the remark following Theorem 4-11, T is surjective if and only if $\dim(\mathrm{Im}(T)) = n$, the dimension of W. Hence it suffices to show that $\dim(\mathrm{Ker}(T)) = 0$ if and only if $\dim(\mathrm{Im}(T)) = n$. But in light of the equality $\dim(\mathrm{Ker}(T)) + \dim(\mathrm{Im}(T)) = n$, this is a triviality, and the proof is complete. $\square$

4-15 COROLLARY. *Let $T: V \to W$ be a linear transformation of finite-dimensional vector spaces. If T is injective, then $\dim(V) \leq \dim(W)$. If T is surjective, then $\dim(V) \geq \dim(W)$.*

PROOF. An easy exercise in the Rank-Nullity Theorem. $\square$

Exercises

1. Let $v_1, \ldots, v_n$ be a linearly independent family in a vector space V. Show that if $i \neq j$, then $v_i \neq v_j$. In other words, a linearly independent family can never contain a repeated vector.

2. Show that the following vectors are linearly independent in $\mathbf{R}^3$:

$$(1,1,1) \text{ and } (0,2,5)$$

3. Show that the following functions are linearly independent in $\mathscr{C}^0(\mathbf{R})$:

$$\sin(x) \text{ and } \cos(x)$$

[Hint: Suppose that there exist $a,b \in \mathbf{R}$ such that $a \cdot \sin(x) + b \cdot \cos(x) = 0$ for all x. Evaluate this linear combination at both $x=0$ and $x=\pi/2$ to determine both a and b.]

4. Give an example of a basis for $\mathbf{R}^2$ other than the canonical basis.

5. Show that each vector $x \in \mathbf{R}^3$ can be expressed uniquely as a linear combination of the following vectors:

$$a=(1,1,1), \ b=(-1,1,0), \ c=(2,0,0)$$

Conclude that a, b, and c constitute a basis for $\mathbf{R}^3$.

6. Let V be the vector space of real polynomials in the indeterminate x of degree less than or equal to 2. Given that the polynomials

$$1, \ 1+x, \text{ and } 1-x^2$$

constitute a basis B for V, find the coordinates of the following polynomial relative to this basis:

$$1-2x+5x^2$$

In other words, compute γ_B for this polynomial.

7. Let V be the subspace of $\mathscr{C}^0(\mathbf{R})$ spanned by the functions e^x and e^{2x}. Show that these functions constitute a basis for V. What is the value of the associated coordinate map for the function $-2e^x + 5e^{2x}$? (Remember that this is a vector in $\mathbf{R}^2$.)

8. Let a_1, a_2, a_3, a_4 be nonzero real numbers. Show that the following set of vectors constitutes a basis B for $\mathbf{R}^4$:

$$(a_1,0,0,0), \ (a_1,a_2,0,0), \ (a_1,a_2,a_3,0), \ (a_1,a_2,a_3,a_4)$$

Find a general formula for the effect of the coordinate map γ_B on a vector $(x_1,x_2,x_3,x_4) \in \mathbf{R}^4$. (This yields a modest example of a *lower triangular* system of equations; note how easily it is solved. Much more on this follows in Section 5.3 below.)

9. Extend the following linearly independent set to a basis for $\mathbf{R}^3$:

$$(1,0,1) \text{ and } (1,1,0)$$

Be sure to establish that your answer indeed constitutes a basis.

10. Give two examples of a real vector space of dimension 4.

11. Give an example of an infinite-dimensional real vector space.

12. Given that the solution space to the differential equation

$$y'' - 2y' + y = 0$$

is a subspace of $\mathscr{C}^2(\mathbf{R})$ of dimension 2, show that the functions

$$e^x \text{ and } xe^x$$

constitute a basis for this subspace. What, then, is the general solution to this equation?

13. Let $T: V \rightarrow V'$ be a linear transformation of real vector spaces. Show that the solution set to the equation $T(v) = 0$ consists of either a single element or infinitely many elements, according to the dimension of the kernel of T.

14. Let $T: V \rightarrow V'$ be a linear transformation of real vector spaces and let v' be an arbitrary element of V'. Show that the solution set to the equation $T(v) = v'$ is either empty, or consists of a single element, or consists of infinitely many elements.

15. Let $T: V \rightarrow V'$ be a surjective linear transformation. Show that if $v_1, \dots, v_n$ span V, then $T(v_1), \dots, T(v_n)$ span V'. In other words, a surjective linear transformation maps a spanning set to a spanning set. [*Hint*: Choose any element $v' \in V'$. It is equal to $T(v)$ for some v in V, since T is surjective. Now express v as a linear combination of the given spanning set; map it over by T and see what happens.]

16. Let $T: V \rightarrow V'$ be an injective linear transformation. Show that if $v_1, \dots, v_n$ is a linearly independent family in V, then $T(v_1), \dots, T(v_n)$ is a linearly independent family in V'. In other words, an injective linear transformation maps an independent set to an independent set. [*Hint*: This is quite elegant. Show that a linear combination of the $T(v_j)$ which equals 0 implies a linear combination of the v_j which lies in the kernel of T. But what is the kernel of an injective linear transformation?]

17. Let $T:V\to V'$ be an isomorphism of vector spaces. Show that if $v_1,\dots,v_n$ is a basis for V, then $T(v_1),\dots,T(v_n)$ is a basis for V'. In other words, an isomorphism maps a basis to a basis. (*Hint*: Read the two previous problems again.)

18. Let V be a finitely-generated vector space over k and let S be a (possibly infinite) spanning set for V. Show that there exists a finite subset of S that also spans V. This justifies one of the preliminary remarks in the proof of Theorem 4-8. [*Hint*: By assumption, V admits some finite spanning set T, every element of which can be written as a (finite!) linear combination of elements of S. Argue from this that only the vectors in S needed for these special linear combinations are required to span V.]

19. Suppose that T is a linear transformation from a vector space of dimension 3 to a vector space of dimension 2. Use the Rank-Nullity Theorem to show that T is not injective.

20. Suppose that T is a linear transformation from a vector space of dimension 3 to a vector space of dimension 4. Use the Rank-Nullity Theorem to show that T is not surjective.

21. Let V be the vector space of real polynomials of degree less than or equal to 2 in the indeterminate x, and consider the linear transformation

$$T:V \;\to\; V$$

$$p \mapsto \frac{d^2 p}{dx^2} - p$$

Show that the kernel of T is trivial. Deduce from this that the map T is an isomorphism. (*Hint*: Directly compute the effect of T on the general quadratic polynomial ax^2+bx+c.)

22. Let $T:V\to W$ be a linear transformation and assume that $\dim(V)=6$ while $\dim(W)=4$. What are the possible dimensions for $\mathrm{Ker}(T)$? Can T be injective? Why or why not?

23. Let V be the vector space of real polynomials of degree less than or equal to 2. Define $T:V\to \mathbf{R}$ by

$$T(p) = \int_{-1}^{+1} p(x)dx$$

Show that T is linear. What is the dimension of the kernel of T? (*Hint*: You need not compute the kernel to do this.)

24. Let V and V' be finite-dimensional vector spaces over a common field and suppose that $\dim(V) \geq \dim(V')$. Show that there exists a surjective linear transformation from V to V'.

25. Let V be a vector space with finite-dimensional subspaces W_0 and W_1 such that $V = W_0 \oplus W_1$. (Review the internal direct sum in Section 3.3 above.) Suppose further that $v_1, \ldots, v_n$ is a basis for W_0 and $u_1, \ldots, u_m$ is a basis for W_1. Show that V is likewise finite-dimensional and that $v_1, \ldots, v_n, u_1, \ldots, u_m$ is a basis for V. [This shows, moreover, that $\dim(V) = \dim(W_0) + \dim(W_1)$.]

26. Let V be a finite-dimensional vector space over a field k and assume that V has basis $v_1, \ldots, v_n$. Show that

$$V = kv_1 \oplus kv_2 \oplus \cdots \oplus kv_n$$

where kv_j denotes the subspace spanned by the single vector v_j in V.

27. Let V be a vector space of dimension 1 over a field k and choose a fixed nonzero element $v_0 \in V$, which is therefore a basis. Let W be any vector space over k and let $w_0 \in W$ be an arbitrary vector. Show that there is a unique linear transformation $T : V \to W$ such that $T(v_0) = w_0$. [*Hint:* What must $T(\lambda v_0)$ be?]

28. Let V be a finite-dimensional vector space over k and let W be any vector space over k. Suppose that $v_1, \ldots, v_n$ is a basis for V and that $w_1, \ldots, w_n$ is an arbitrary family in W. Use the two previous problems and the universal property of internal direct sums to show that there exists a unique linear transformation $T : V \to W$ such that $T(v_j) = w_j$ $(j = 1, \ldots, n)$. (We shall revisit this important result in Chapter 6, giving a less abstract proof.)

29. Let V be a finite-dimensional vector space and assume that W_0 is a subspace of V. Show that there exists a complementary subspace W_1 of V such that $V = W_0 \oplus W_1$. (*Hint:* Extend a basis for W_0 to a basis for V.)

30. Let $T : V \to V'$ be a surjective linear transformation of finite-dimensional vector spaces. Show that there exists a subspace W of V such that

$$V = \mathrm{Ker}(T) \oplus W \quad \text{with} \quad W \cong V'$$

(*Hint:* Use the proof of the Rank-Nullity Theorem.)

31. Give an example of an infinite-dimensional vector space V and a linear transformation $T: V \to V$ such that both $\text{Im}(T)$ and $\text{Ker}(T)$ are also infinite dimensional. (*Hint*: Try $V = \mathbf{R}[x]$, the vector space of polynomials with real coefficients.)

5
Matrices

This chapter formally introduces matrices and matrix algebra. First, we take a superficial look at matrices simply as arrays of scalars, divorced from any extrinsic meaning. We define the elements of matrix arithmetic and show that it is formally well behaved, which is surprising since the definition of matrix multiplication looks somewhat unnatural at this point. Second, we consider the connection between matrices and linear systems of equations. This begins to build a bridge between vector space theory and matrix theory that will be completed in the following chapter. Finally, we survey two powerful solution techniques for linear systems and a related method for matrix inversion.

5.1 Notation and Terminology

DEFINITION. Let k be a field. Then an *$m \times n$ matrix over k* is an array of the form

$$A = \begin{pmatrix} a_{11} & a_{12} & \cdots & a_{1n} \\ a_{21} & a_{22} & \cdots & a_{2n} \\ \vdots & \vdots & & \vdots \\ a_{m1} & a_{m2} & \cdots & a_{mn} \end{pmatrix}$$

where each entry a_{ij} lies in k. We often write $A = (a_{ij})$ to indicate that A is a matrix whose (i,j)-entry is a_{ij}. The set of all such matrices is denoted $\text{Mat}_{m \times n}(k)$.

One finds it convenient to establish notation describing both the rows and columns of A. Henceforth $A^1, \ldots, A^n$ will denote its columns and $A_1, \ldots, A_m$ will denote its rows. These may be regarded as vectors in k^m or k^n. [Indeed, vectors in k^n may likewise be considered either $1 \times n$ matrices (*row matrices*) or $n \times 1$ matrices (*column matrices*). The latter interpretation is more prominent.] We often consider A to be the amalgamation of its columns and write

$$A = (A^1, \ldots, A^n)$$

In the important special case that $m=n$, we speak of *square matrices* and let $M_n(k)$ denote the set of all square matrices with entries in k.

We now introduce two binary operations and an external scalar operation on matrices. Addition and scalar multiplication are natural and obvious; multiplication is natural but obscure.

DEFINITIONS. Let $A=(a_{ij})$ and $B=(b_{ij})$ lie in $\mathrm{Mat}_{m\times n}(k)$. Then their sum $A+B$ is the $m \times n$ matrix $C=(c_{ij})$ defined by

$$c_{ij} = a_{ij} + b_{ij} \quad (1 \le i \le m,\ 1 \le j \le n)$$

For any $\lambda \in k$, the scalar product λA is the $m \times n$ matrix whose (i,j)-component is λa_{ij}; that is, $\lambda A = (\lambda a_{ij})$.

Thus addition is only defined for matrices of the same size, in which case it amounts to addition of corresponding components. Scalar multiplication has the effect of multiplying every entry by the given scalar. These operations are remarkably similar to those defined on k^n, and in fact we have the following fundamental result.

5-1 PROPOSITION. *For all m and n, $\mathrm{Mat}_{m\times n}(k)$ is a vector space over k of dimension mn. In particular, the matrices of a given size form an additive group with respect to matrix addition.*

PROOF. The proof is entirely similar to that for k^n. The zero matrix of a given size is that matrix whose entries are all 0. To see that the dimension of the space $\mathrm{Mat}_{m\times n}(k)$ is mn, note that the following mn matrices constitute a basis:

E_{ij} = the matrix whose (i,j)-component is 1, with all other components 0 $(1 \le i \le m,\ 1 \le j \le n)$ □

We now move on to the more interesting operation—matrix multiplication.

DEFINITION. Let $A \in \mathrm{Mat}_{m\times n}(k)$ and $B \in \mathrm{Mat}_{n\times p}(k)$, so that the number of columns of A is equal to the number of rows of B. Then we define their product AB to be the $m \times p$ matrix $C=(c_{ij})$ defined by

$$c_{ij} = \sum_{k=1}^{n} a_{ik} b_{kj}$$

Note that the variable k used here as a summation index is not to be confused with the field k. (The distinction, in general, is clear from the context.) When expanded, the summation above has the form

$$c_{ij} = a_{i1}b_{1j} + a_{i2}b_{2j} + \cdots + a_{in}b_{nj}$$

from which one sees that the (i,j)-component of AB is the "dot product" of the ith row of A with the jth column of B.

EXAMPLE. We give one numerical example of a matrix product. (Any further such computations would be antithetical to the ethereal spirit of this book!) We calculate the product of a 2×3 matrix with a 3×4 matrix; the result is a 2×4 matrix:

$$\begin{pmatrix} 1 & 1 & 0 \\ 0 & 2 & 1 \end{pmatrix} \begin{pmatrix} 0 & 2 & 0 & 1 \\ 0 & 1 & 1 & 1 \\ 2 & 0 & 2 & 1 \end{pmatrix} = \begin{pmatrix} 0 & 3 & 1 & 2 \\ 2 & 2 & 4 & 3 \end{pmatrix}$$

We review in particular the calculation of the (2,3)-entry of the product. This is the dot product of the row 2 of the left-hand factor with column 3 of the right-hand factor. Hence $0 \cdot 0 + 2 \cdot 1 + 1 \cdot 2 = 4$.

Note that the product of an $m \times n$ matrix with an $n \times 1$ column matrix is defined and results in an $m \times 1$ column matrix. In terms of these special products, one may show easily that

$$AB = (A \cdot B^1, \ldots, A \cdot B^p) \tag{5.1}$$

This is to say that to compute AB we may multiply each column of B by A and then amalgamate the resulting columns.

WARNING. Matrix multiplication is *not* commutative. Worse yet, BA may not be defined, even though AB is, unless A has as many rows as B has columns.

A special matrix in $M_n(k)$ that arises in connection with matrix multiplication is the $n \times n$ *identity matrix*, denoted I_n, whose (i,j)-entry, often denoted δ_{ij}, is 1 if $i=j$ and 0 otherwise. (The symbol δ_{ij} is called the *Kronecker delta*.) Thus

$$I_n = \begin{pmatrix} 1 & 0 & \ldots & 0 \\ 0 & 1 & \ldots & 0 \\ \vdots & \vdots & & \vdots \\ 0 & 0 & \ldots & 1 \end{pmatrix}$$

I_n has the special property that $I_nA=A$ and $AI_n=A$ whenever these products are defined. The identity matrix is a special case of a *diagonal matrix*, a square matrix which has nonzero entries only on its top-left to bottom-right diagonal.

We now summarize the principal arithmetic properties of matrix multiplication. Not surprisingly, $M_n(k)$ fits snugly into one of the abstract structures introduced in Chapter 2.

5-2 PROPOSITION. *Matrix arithmetic has the following properties, whenever the indicated sums and products are defined:*

(i) $(AB)C = A(BC)$

(ii) $A(B+C) = AB + AC$

(iii) $(A+B)C = AC + BC$

(iv) $\lambda(AB) = (\lambda A)B = A(\lambda B)$ $\forall \lambda \in k$

In particular, $M_n(k)$ is a ring with unity.

PROOF. All of these may be proved by direct calculation. We shall only prove (ii), leaving (iii) and (iv) as exercises and postponing (i) until later, when a particularly elegant argument will suggest itself. So let $A = (a_{ij})$ be an $m \times n$ matrix and let $B = (b_{ij})$ and $C = (c_{ij})$ be $n \times p$ matrices. We must show that corresponding components of $A(B+C)$ and $AB + AC$ match. The (i,j)-entry of $B+C$ is $b_{ij} + c_{ij}$, whence the (i,j)-entry of $A(B+C)$ is by definition

$$\sum_{k=1}^{n} a_{ik}(b_{kj} + c_{kj}) = \sum_{k=1}^{n} (a_{ik}b_{kj} + a_{ik}c_{kj}) = \sum_{k=1}^{n} a_{ik}b_{kj} + \sum_{k=1}^{n} a_{ik}c_{kj}$$

Here we have freely used the arithmetic properties of the field k. The right-hand side of the equation is precisely the sum of the (i,j)-entry of AB with the (i,j)-entry of AC, and this is evidently the (i,j)-entry of $AB + AC$, thus completing the proof. $\square$

We conclude this section with two other basic notions: the transpose and the inverse. (The transpose is of far greater importance than is suggested by the following definition. Chapter 6 provides enlightenment.)

DEFINITION. Let $A = (a_{ij})$ be an $m \times n$ matrix. Then ^tA, the *transpose of A*, is the $n \times m$ matrix whose (i,j)-component is a_{ji}; that is, $^tA = (a_{ji})$. A (necessarily square) matrix equal to its own transpose is called *symmetric.*

More conceptually, we obtain the transpose of A from A itself by interchanging its rows and columns. Clearly a diagonal matrix is symmetric.

EXAMPLE. We transpose a 2×3 matrix to obtain a 3×2 matrix:

$$
{}^t\!\begin{pmatrix} 1 & 0 & 2 \\ 0 & 2 & 1 \end{pmatrix} = \begin{pmatrix} 1 & 0 \\ 0 & 2 \\ 2 & 1 \end{pmatrix}
$$

Matrix transposition is linear, but not quite multiplicative, as this next result shows.

5-3 PROPOSITION. *The matrix transpose has the following properties, whenever the indicated sums and products are defined:*

(i) ${}^t(A+B) = {}^t\!A + {}^t\!B$

(ii) ${}^t(\lambda A) = \lambda\,{}^t\!A \quad \forall \lambda \in k$

(iii) ${}^t(AB) = {}^t\!B\,{}^t\!A$

We leave the proofs as exercises. A noncomputational explanation of (iii) follows easily from the results of Section 6.4. (See Chapter 6, Exercise 26.) ❑

DEFINITION. An element A of $M_n(k)$ is called *invertible* or *nonsingular* if there exists an element $B \in M_n(k)$ such that $AB = I_n = BA$. The set of all such invertible matrices is denoted $GL_n(k)$ and called the *general linear group of rank n matrices over k.*

5-2a PROPOSITION. $GL_n(k)$ *is a group for all n and noncommutative for $n > 1$.*

PROOF. The student should show that for any ring A with unity, the set of multiplicatively invertible elements of A constitutes a group under the ring multiplication. (Closure is the only issue, albeit a small one.) This is called the *group of units* of A and denoted $A^\times$. By definition, $GL_n(k)$ is the group of units in the matrix ring $M_n(k)$, and hence a group. Noncommutativity is slightly more subtle, but one shows easily that the following calculation in $GL_2(k)$, which works over any field, can be "imported" into any larger matrix ring:

$$
\begin{pmatrix} 0 & 1 \\ 1 & 0 \end{pmatrix}\begin{pmatrix} 1 & 0 \\ 1 & 1 \end{pmatrix} = \begin{pmatrix} 1 & 1 \\ 1 & 0 \end{pmatrix}
$$

$$
\begin{pmatrix} 1 & 0 \\ 1 & 1 \end{pmatrix}\begin{pmatrix} 0 & 1 \\ 1 & 0 \end{pmatrix} = \begin{pmatrix} 0 & 1 \\ 1 & 1 \end{pmatrix}
$$

❑

The proposition also shows that the ring $M_n(k)$ is noncommutative for $n > 1$.

The group $GL_n(k)$ is one of the most fundamental objects in mathematics, for reasons which will begin to emerge in the following section. Indeed, linear algebra abounds with tests for invertibility.

5.2 Introduction to Linear Systems

Throughout this section we regard vectors in k^n as columns (i.e., $n \times 1$ matrices). Let $A = (a_{ij})$ be an $m \times n$ matrix with entries in k. Then given $x \in k^n$ the product Ax lies in k^m. Hence we have a mapping

$$k^n \rightarrow k^m$$
$$x \mapsto Ax$$

which by the basic properties of matrix arithmetic (Proposition 5-2) is a linear transformation:

$$A(x+y) = Ax + Ay \quad \text{and} \quad A(\lambda x) = \lambda \cdot Ax$$

Let us explicitly describe Ax in terms of its components:

$$\begin{pmatrix} a_{11} & a_{12} & \cdots & a_{1n} \\ a_{21} & a_{22} & \cdots & a_{2n} \\ \vdots & \vdots & & \vdots \\ a_{m1} & a_{m2} & \cdots & a_{mn} \end{pmatrix} \begin{pmatrix} x_1 \\ x_2 \\ \vdots \\ x_n \end{pmatrix} = \begin{pmatrix} a_{11}x_1 + a_{12}x_2 + \cdots + a_{1n}x_n \\ a_{21}x_1 + a_{22}x_2 + \cdots + a_{2n}x_n \\ \vdots \\ a_{m1}x_1 + a_{m2}x_2 + \cdots + a_{mn}x_n \end{pmatrix}$$

We record two results apparent from this imposing equation.

(1) The matrix product Ae_j of A with the jth canonical basis vector for k^n is precisely the jth column of A. (In the explicit equation above, x_j is 1 while all other x's are 0. Hence only the terms corresponding to the jth column appear on the right.) That is,

$$Ae_j = A^j \tag{5.2}$$

This simple formula appears repeatedly at critical points throughout the remainder of the text.

(2) The linear system of m equations in n unknowns

$$a_{11}x_1 + \cdots + a_{1n}x_n = y_1$$
$$a_{21}x_1 + \cdots + a_{2n}x_n = y_2$$
$$\vdots$$
$$a_{m1}x_1 + \cdots + a_{mn}x_n = y_m$$

is equivalent to the single matrix equation $Ax=y$. This might also be expressed as a vector equation as follows:

$$\begin{pmatrix} a_{11} \\ a_{21} \\ \vdots \\ a_{m1} \end{pmatrix} x_1 + \cdots + \begin{pmatrix} a_{1n} \\ a_{2n} \\ \vdots \\ a_{mn} \end{pmatrix} x_n = \begin{pmatrix} y_1 \\ y_2 \\ \vdots \\ y_m \end{pmatrix}$$

In the special case that $y=0$ (so that all of the right-hand terms are 0), we speak of a *homogeneous linear system*. Note that the scalars are written on the right—an acceptable abuse in this one case.

We now record some fundamental results about linear systems which follow directly from our results on linear transformations and dimension. Henceforth, if A is an $m \times n$ matrix over k, we let T_A denote the corresponding linear transformation from k^n into k^m which sends x to Ax; that is,

$$T_A : k^n \to k^m$$
$$x \mapsto Ax$$

The *rank* of A, denoted rk(A), is the rank of the linear transformation T_A (the dimension of its image). From the vector form of $Ax=y$ shown above, this just amounts to the dimension of the subspace spanned by the columns of A. The *nullity* of A is the nullity of T_A (the dimension of its kernel). Note that Ker(T_A) is exactly the solution set to the homogeneous linear system $Ax=0$.

5-4 PROPOSITION. *The set of all y for which there exists a solution to the linear system $Ax=y$ of m equations in n unknowns is a subspace of k^m. The solution set to the homogeneous system $Ax=0$ is likewise a subspace of k^n.*

PROOF. The set of all y in k^m for which there exists a solution to the given linear system is precisely the image of T_A. Similarly, as remarked above, the solution set to the homogeneous equation $Ax=0$ is just the kernel of T_A. But we know from Section 3.2 that both the image and the kernel of a linear transformation are subspaces of their ambient spaces.   □

5-5 PROPOSITION. *Suppose that $n > m$. Then the homogeneous linear system of m equations in n unknowns represented by the matrix equation $Ax = 0$ always has a nontrivial (i.e., nonzero) solution. In fact, the solution set is a subspace of k^n of dimension at least $n - m$.*

PROOF. The rank of A, being the dimension of $\text{Im}(T_A)$, a subspace of k^m, is necessarily less than or equal to m and therefore strictly less than n. Since rank plus nullity must equal n, this implies that the nullity of A [the dimension of $\text{Ker}(T_A)$] is positive and at least $n - m$. But this says precisely that the solution space to the homogeneous system $Ax = 0$ has dimension at least $n - m$, just as claimed.   □

We now consider a case of utmost importance: n equations in n unknowns.

5-6 THEOREM. *Let $A \in M_n(k)$. Then the following six statements are equivalent:*

(i)   *The linear system $Ax = y$ has at least one solution for all $y \in k^n$.*

(ii) *The columns of A span k^n.*

(iii) *The homogeneous system $Ax = 0$ has only the trivial solution $x = 0$.*

(iv) *The columns of A are linearly independent.*

(v)   *The linear system $Ax = y$ has exactly one solution for all $y \in k^n$.*

(vi) *The columns of A constitute a basis for k^n.*

Moreover, a sufficient condition for any, hence all, of these statements is that the matrix A be invertible; i.e., $A \in GL_n(k)$.

Later we shall strengthen this theorem to see, in particular, that the invertibility of A is not only sufficient but necessary.

PROOF. We have already seen in Theorem 4-11 that (ii), (iv), and (vi) are equivalent. The equivalences (i) $\Leftrightarrow$ (ii), (iii) $\Leftrightarrow$ (iv) and (v) $\Leftrightarrow$ (vi) are all apparent from the representation of the given linear system in vector form:

$$\begin{pmatrix} a_{11} \\ a_{21} \\ \vdots \\ a_{n1} \end{pmatrix} x_1 + \cdots + \begin{pmatrix} a_{1n} \\ a_{2n} \\ \vdots \\ a_{nn} \end{pmatrix} x_n = \begin{pmatrix} y_1 \\ y_2 \\ \vdots \\ y_n \end{pmatrix}$$

This establishes the equivalence of the six statements. We observe next that the invertibility of A implies (v). This is trivial: if A is invertible, then the linear system $Ax=y$ is solved uniquely by multiplying both sides of the equation by A^{-1} to obtain $x=A^{-1}y$. This completes the proof.   □

It is tempting at this point to establish the necessity of the invertibility of A as follows. Let B^j (here viewed as a column vector in k^n) be the unique solution to $Ax=e_j$ which is assumed to exist by (v). Let $B=(B^1,...,B^n)$ be the amalgamation of these columns into an $n \times n$ matrix. Then according to Eq. 5.1,

$$AB = A(B^1,...,B^n) = (AB^1,...,AB^n) = (e_1,...,e_n) = I_n$$

and B is at least a right inverse for A. The problem is that since we have not yet established that A and B lie in the group $GL_n(k)$, we cannot apply elementary group theory to conclude that B is in fact A^{-1}. What we need to show is that we can construct a left inverse for A in a similar manner, and for this we need to know that the rows and columns of A span subspaces of equal dimension. This key result is better proved in the following chapter after we have a deeper understanding of the connection between matrices and linear transformations.

5.3 Solution Techniques

We shall now survey two powerful and popular techniques for solving the linear system $Ax=y$. Note that in general there may be none, one, or many solutions, and the matrix A need not be square.

REMARKS

(1) If A is square and invertible, we of course have the option of finding A^{-1} and solving $Ax=y$ by multiplying both sides by A^{-1}, as in the proof of the previous theorem. This is usually not efficient, however, since the problem of inverting A is often far more difficult than a direct solution to the original equation.

(2) For any but the smallest systems, solutions are almost invariably found via computer using any of the widely available numerical packages. (Hand calculations are often intractable and at the very least invite miscalculation; they should only be trusted when thoroughly checked.) Since computer arithmetic is usually only approximate, one must then be concerned with several subtle numerical analytic issues that can dramatically affect the final outcome. Such considerations are beyond the scope of our discussion and we refer the reader to texts on numerical methods or numerical analysis.

Elimination Methods

First note that the linear system

$$a_{11}x_1 + \cdots + a_{1n}x_n = y_1$$
$$a_{21}x_1 + \cdots + a_{2n}x_n = y_2$$
$$\vdots$$
$$a_{m1}x_1 + \cdots + a_{mn}x_n = y_m$$

may be represented by the *augmented matrix*

$$\left(\begin{array}{cccc|c} a_{11} & a_{12} & \cdots & a_{1n} & y_1 \\ a_{21} & a_{22} & \cdots & a_{2n} & y_2 \\ \vdots & \vdots & & \vdots & \vdots \\ a_{m1} & a_{m2} & \cdots & a_{mn} & y_m \end{array} \right)$$

which retains all of the information of the original system. [We often condense this to $(A\,|\,y)$, when convenient.] One can obviously apply the following transformations to the augmented matrix without disturbing the solutions to the associated system:

(i) Interchanging two rows. (We represent the interchange of row i with row j by $R_i \leftrightarrow R_j$.)

(ii) Multiplication of a row by a nonzero scalar. (We represent the multiplication of row i by a constant λ by $R_i \leftarrow \lambda R_i$.)

(iii) Addition of a scalar multiple of one row to another. (We represent the addition of λ times row j to row i by $R_i \leftarrow R_i + \lambda R_j$.)

These are called *elementary row operations* and represent familiar transformations routinely applied to systems of equations. Their particular value with respect to linear systems lies in the elimination of variables.

EXAMPLE. Consider the system

$$x_1 - x_2 = 4$$
$$2x_1 - x_2 = 7$$

We apply the following set of elementary row operations to the augmented matrix:

$$\begin{pmatrix} 1 & -1 & | & 4 \\ 2 & -1 & | & 7 \end{pmatrix} \xrightarrow[R_2 \leftarrow R_2 - 2R_1]{} \begin{pmatrix} 1 & -1 & | & 4 \\ 0 & 1 & | & -1 \end{pmatrix} \xrightarrow[R_1 \leftarrow R_1 + R_2]{} \begin{pmatrix} 1 & 0 & | & 3 \\ 0 & 1 & | & -1 \end{pmatrix}$$

Reconstituting the last matrix as a linear system, we have $x_1 = 3$ and $x_2 = -1$.

Of course, the example represents an unreasonably simple case, but it does manifest the general nature of the *Gauss-Jordan elimination method*, which we now sketch. The basic idea is to apply the elementary row operations to $(A | y)$ to bring the matrix A into conformance with four conditions:

(i)   all (if any) rows of A consisting entirely of 0's lie at the bottom of A;

(ii) the first nonzero entry of each nonzero row is 1;

(iii) the leading 1's of the nonzero rows move strictly to the right as we proceed down the matrix;

(iv) all entries above and below the leading 1 of a nonzero row are 0.

If conditions (i)-(iii) are satisfied, we say that A is in *row-echelon form*. If conditions (i)-(iv) are satisfied, we say that A is in *reduced row-echelon form*. One can show by induction that every matrix can be brought to this latter form by elementary row operations. Note that whatever transformations are applied to A are also applied to y.

We shall now give several examples of augmented matrices in reduced row-echelon form both to elucidate the conditions above and to show that a system in this reduced form is easily solved.

EXAMPLES

(1) Consider a 4×4 system whose augmented matrix in reduced row-echelon form is

$$\begin{pmatrix} 1 & 0 & 0 & 0 & | & 5 \\ 0 & 1 & 0 & 0 & | & 1 \\ 0 & 0 & 1 & 0 & | & 0 \\ 0 & 0 & 0 & 1 & | & 4 \end{pmatrix}$$

This is the simplest case of all—we have reached the identity matrix. The solution to the original system is evidently $x_1 = 5$, $x_2 = 1$, $x_3 = 0$, $x_4 = 4$. Similar considerations apply to any $n \times n$ system whose reduced row-echelon form amounts to I_n: the solution may be read off directly from the right column.

(2) Consider a 3×3 system whose augmented matrix in reduced row-echelon form is

$$\begin{pmatrix} 1 & 0 & 0 & | & 2 \\ 0 & 0 & 1 & | & 7 \\ 0 & 0 & 0 & | & 0 \end{pmatrix}$$

The general solution is $x_1 = 2$, $x_3 = 7$, with no conditions whatsoever on x_2. In fact, whenever a column in reduced form consists entirely of 0's, the corresponding variable is left unconstrained and plays no further role in the solution.

(3) Consider a 4×5 system whose augmented matrix in reduced row-echelon form is

$$\begin{pmatrix} 1 & 2 & 0 & 0 & 0 & | & 1 \\ 0 & 0 & 1 & 0 & 0 & | & 0 \\ 0 & 0 & 0 & 1 & 0 & | & 2 \\ 0 & 0 & 0 & 0 & 0 & | & 4 \end{pmatrix}$$

We see that there are no solutions since the last row would otherwise imply that $0 = 4$. We say that the system is *inconsistent*. This occurs whenever we observe a row for which every entry but the last is 0. (More fundamentally, the system $Ax = y$ is inconsistent if and only if y does not lie in the image of the associated linear transformation T_A.)

(4) Consider the 4×5 system whose augmented matrix in reduced row-echelon form is

$$\begin{pmatrix} 1 & 2 & 0 & 0 & -1 & | & 1 \\ 0 & 0 & 1 & 0 & -4 & | & 5 \\ 0 & 0 & 0 & 1 & 6 & | & 2 \\ 0 & 0 & 0 & 0 & 0 & | & 0 \end{pmatrix}$$

This system is not inconsistent and generalizes the situation in Example 2 insofar as the variables corresponding to columns which do not manifest a leading 1 (in this case columns 2 and 5) are unconstrained. These variables, however, do play a further role in the general solution, which might be expressed thus:

$$x_1 = 1 - 2x_2 + x_5$$
$$x_3 = 5 + 4x_5$$
$$x_4 = 2 - 6x_5$$
$$x_2, x_5 \text{ arbitrary}$$

This concludes our introduction to elimination techniques. We now pass on to a second approach—one with a decidedly distinct flavor.

LU Decomposition

This discussion applies to $n \times n$ systems. First, a matrix $L = (a_{ij})$ is called *lower triangular* if $a_{ij} = 0$ whenever $i < j$. A matrix $U = (b_{ij})$ is called *upper triangular* if $b_{ij} = 0$ whenever $j < i$. Thus

$$
L = \begin{pmatrix}
a_{11} & 0 & 0 & \cdots & 0 \\
a_{21} & a_{22} & 0 & \cdots & 0 \\
a_{31} & a_{32} & a_{33} & \cdots & 0 \\
\vdots & \vdots & \vdots & & \vdots \\
a_{n1} & a_{n2} & a_{n3} & \cdots & a_{nn}
\end{pmatrix}
\qquad
U = \begin{pmatrix}
b_{11} & b_{12} & b_{13} & \cdots & b_{1n} \\
0 & b_{22} & b_{23} & \cdots & b_{2n} \\
0 & 0 & b_{33} & \cdots & b_{3n} \\
\vdots & \vdots & \vdots & & \vdots \\
0 & 0 & 0 & \cdots & b_{nn}
\end{pmatrix}
$$

The beauty of triangular matrices in connection with linear systems is that equations such as $Lx = y$ and $Ux = y$ are easily solved by *forward* and *back substitution*, respectively. To be precise, the lower triangular system

$$
\begin{aligned}
a_{11}x_1 &= y_1 \\
a_{21}x_1 + a_{22}x_2 &= y_2 \\
&\vdots \\
a_{n1}x_1 + a_{n2}x_2 + \cdots + a_{nn}x_n &= y_n
\end{aligned}
$$

has solution

$$
\begin{aligned}
x_1 &= y_1 / a_{11} \\
x_2 &= (y_2 - a_{21}x_1) / a_{22} \\
&\vdots \\
x_n &= (y_n - a_{n1}x_1 - a_{n2}x_2 - \cdots - a_{n,n-1}x_{n-1}) / a_{nn}
\end{aligned}
$$

(provided none of the diagonal entries is 0), and this is numerically trivial to compute. A similar set of equations applies to an upper triangular system with the substitutions running from the bottom upward. This leads to a most elegant solution technique for the general nonsingular $n \times n$ system $Ax = y$, as follows.

Suppose that we can find, respectively, lower and upper triangular matrices L and U such that $A = LU$. Then with this factorization in hand, we solve the linear system $Ax = y$ in two steps:

(i) Solve $Lz = y$ for z.

(ii) Solve $Ux = z$ for x.

As described above, both solutions are straightforward, and we can see at once that the vector x thus obtained solves the original system:

$$Ax = (LU)x = L(Ux) = Lz = y$$

This method is effective in practice because there is a numerically efficient technique known as *Crout's algorithm* for factoring A into the product of a lower and an upper triangular matrix. We shall not describe Crout's algorithm here, but it can be found in almost any book on numerical methods.

EXAMPLE. Given the factorization

$$A = \begin{pmatrix} 2 & 1 & 0 \\ 4 & 1 & 1 \\ 2 & -3 & 2 \end{pmatrix} = \begin{pmatrix} 1 & 0 & 0 \\ 2 & 1 & 0 \\ 1 & 4 & 1 \end{pmatrix} \begin{pmatrix} 2 & 1 & 0 \\ 0 & -1 & 1 \\ 0 & 0 & -2 \end{pmatrix}$$

we solve the linear system

$$\begin{pmatrix} 2 & 1 & 0 \\ 4 & 1 & 1 \\ 2 & -3 & 2 \end{pmatrix} \begin{pmatrix} x_1 \\ x_2 \\ x_3 \end{pmatrix} = \begin{pmatrix} 3 \\ 7 \\ 3 \end{pmatrix}$$

using the two-step procedure described above. First, by forward substitution we solve

$$\begin{pmatrix} 1 & 0 & 0 \\ 2 & 1 & 0 \\ 1 & 4 & 1 \end{pmatrix} \begin{pmatrix} z_1 \\ z_2 \\ z_3 \end{pmatrix} = \begin{pmatrix} 3 \\ 7 \\ 3 \end{pmatrix}$$

to obtain $z_1 = 3$, $z_2 = 1$, $z_3 = -4$. Next solve

$$\begin{pmatrix} 2 & 1 & 0 \\ 0 & -1 & 1 \\ 0 & 0 & -2 \end{pmatrix} \begin{pmatrix} x_1 \\ x_2 \\ x_3 \end{pmatrix} = \begin{pmatrix} 3 \\ 1 \\ -4 \end{pmatrix}$$

by back substitution to obtain $x_1 = 1$, $x_2 = 1$, $x_3 = 2$.

5.4 Multiple Systems and Matrix Inversion

We elaborate briefly on a feature implicit in the Gauss-Jordan elimination method introduced in the previous section. Suppose that given a fixed $m \times n$ matrix A and vectors $y_1, y_2 \in k^n$, we are presented with two linear systems

$$Ax_1 = y_1 \quad \text{and} \quad Ax_2 = y_2$$

the point being to find respective solutions x_1 and x_2 to each. Observe that the sequence of elementary row operations required to bring A to reduced row-echelon form is entirely determined by A, so that in this sense the data on the right participate rather passively. Therefore the particular sequence of operations needed to solve $Ax_1 = y_1$ will be precisely the same as that required to solve $Ax_2 = y_2$, the only difference being the arithmetic that takes place in the right-most column of the augmented matrix. Thus an efficient method of solution is to represent the two systems by the (doubly) augmented matrix

$$\begin{pmatrix} a_{11} & a_{12} & \cdots & a_{1n} & y_{11} & y_{12} \\ a_{21} & a_{22} & \cdots & a_{2n} & y_{21} & y_{22} \\ \vdots & \vdots & & \vdots & \vdots & \vdots \\ a_{m1} & a_{m2} & \cdots & a_{mn} & y_{m1} & y_{m2} \end{pmatrix}$$

and then to row-reduce A. We can then read both solutions (if any) off the reduced augmented matrix, as previously.

EXAMPLE. We revisit the system

$$x_1 - x_2 = 4$$
$$2x_1 - x_2 = 7$$

and simultaneously consider the second system (with the same coefficients)

$$x_1 - x_2 = 2$$
$$2x_1 - x_2 = 1$$

We apply the same sequence of elementary row operations as previously to the doubly augmented matrix representing both data sets:

$$
\begin{pmatrix} 1 & -1 & | & 4 & 2 \\ 2 & -1 & | & 7 & 1 \end{pmatrix} \xrightarrow[R_2 \leftarrow R_2 - 2R_1]{} \begin{pmatrix} 1 & -1 & | & 4 & 2 \\ 0 & 1 & | & -1 & -3 \end{pmatrix}
$$

$$
\xrightarrow[R_1 \leftarrow R_1 + R_2]{} \begin{pmatrix} 1 & 0 & | & 3 & -1 \\ 0 & 1 & | & -1 & -3 \end{pmatrix}
$$

The respective solutions are thus $x_1 = (3,-1)$ and $x_2 = (-1,-3)$. Note that this is only slightly more work than solving the original single system.

These same considerations apply more generally to r systems of the form

$$
Ax_j = y_j \quad (j=1,\ldots,r)
$$

for any positive r. This is evidently equivalent (by Eq. 5.1) to the single matrix equation

$$
AX = Y \tag{5.3}
$$

where $X = (x_1,\ldots,x_r)$ is an $n \times r$ matrix and $Y = (y_1,\ldots,y_r)$ is an $m \times r$ matrix. For computational purposes, we may represent Eq. 5.3 by the augmented matrix $(A|Y)$. To solve it, we row-reduce A and read the solutions (if any) off the transformed augmented matrix. (As in the single-system case, we obtain a unique solution if and only if A reduces to the identity matrix.)

For $A \in M_n(k)$, a special case of utmost importance is the equation

$$
AX = I_n
$$

the solution to which yields A^{-1}, if in fact A is invertible. (See the remarks following the proof of Theorem 5-6.) Thus we have established a recipe for matrix inversion that amounts to the following:

Apply elementary row operations to the augmented matrix $(A|I_n)$ to obtain, if possible, $(I_n|X)$. At this point, $A^{-1} = X$.

EXAMPLE. To illustrate, we shall invert the matrix of coefficients from our previous example:

$$
\begin{pmatrix} 1 & -1 \\ 2 & -1 \end{pmatrix}
$$

The calculation proceeds as follows:

$$\begin{pmatrix} 1 & -1 & \vdots & 1 & 0 \\ 2 & -1 & \vdots & 0 & 1 \end{pmatrix} \xrightarrow[R_2 \leftarrow R_2 - 2R_1]{} \begin{pmatrix} 1 & -1 & \vdots & 1 & 0 \\ 0 & 1 & \vdots & -2 & 1 \end{pmatrix}$$

$$\xrightarrow[R_1 \leftarrow R_1 + R_2]{} \begin{pmatrix} 1 & 0 & \vdots & -1 & 1 \\ 0 & 1 & \vdots & -2 & 1 \end{pmatrix}$$

Thus

$$\begin{pmatrix} 1 & -1 \\ 2 & -1 \end{pmatrix}^{-1} = \begin{pmatrix} -1 & 1 \\ -2 & 1 \end{pmatrix}$$

as one can easily verify by direct calculation.

REMARK. As described in Exercise 7 below, there is an effective direct formula for the inversion of 2×2 matrices; one would not generally use the present method in such cases.

Exercises

1. Solve the following matrix equation for x, y, z, and w:

$$\begin{pmatrix} 1 & 2 \\ 0 & 1 \end{pmatrix}\begin{pmatrix} x & y \\ z & w \end{pmatrix} = \begin{pmatrix} 10 & 2 \\ 4 & 2 \end{pmatrix}$$

2. Suppose that

$$\begin{pmatrix} 1 & 1 \\ 0 & 1 \end{pmatrix}\begin{pmatrix} x & y \\ z & w \end{pmatrix} = \begin{pmatrix} x & y \\ z & w \end{pmatrix}\begin{pmatrix} 1 & 1 \\ 0 & 1 \end{pmatrix}$$

Show that $x=w$, $z=0$, and there is no constraint on y.

3. Let A be an $m \times n$ matrix. Show that the matrix products $A \cdot {}^tA$ and ${}^tA \cdot A$ are both defined. Compute these products for the matrix

$$A = \begin{pmatrix} 2 & 0 & 1 \\ 0 & 1 & 1 \end{pmatrix}$$

4. What is the dimension of the space of 4×4 symmetric matrices defined over a field k? Of all $n×n$ symmetric matrices? (*Hint*: Count the degrees of freedom.)

5. What is the dimension of the space of all 5×5 upper triangular matrices in $M_n(k)$? Of all $n×n$ upper triangular matrices?

6. Show that the product of two invertible matrices in $M_n(k)$ is invertible.

7. Let

$$A = \begin{pmatrix} a & b \\ c & d \end{pmatrix}$$

lie in $M_2(k)$ and assume that $ad - bc \neq 0$. Show that A^{-1} exists and that in fact

$$A^{-1} = \frac{1}{ad - bc} \begin{pmatrix} d & -b \\ -c & a \end{pmatrix}$$

Conversely, show that if $ad - bc = 0$, then A is not invertible. [*Hints*: To verify the formula for A^{-1}, try computing its product with A. For the converse, show that if $ad-bc=0$, then the columns of A are linearly dependent. Hence A cannot be invertible, according to Theorem 5-6.]

8. Let B be the subset of $GL_2(k)$ consisting of all matrices of the form

$$A = \begin{pmatrix} a & b \\ 0 & c \end{pmatrix}, \quad ac \neq 0$$

(Such matrices are indeed invertible according to the previous problem.) Show that B is a subgroup of $GL_2(k)$.

9. For this problem and the next, superscripts on matrices indicate exponents rather than columns. Thus A^2 means $A \cdot A$, not the second column of A.

Find a 2×2 matrix A such that $A \neq 0$ (that is, A is not the zero matrix) but nonetheless $A^2 = 0$. Next, find a 3×3 matrix such that $A, A^2 \neq 0$, but $A^3 = 0$. Finally, for arbitrary n find an $n×n$ matrix A, such that $A, A^2, A^3, \ldots, A^{n-1} \neq 0$, but $A^n = 0$. (The following chapter will show you how better to think about this problem.)

10. Let $A \in M_n(k)$ be such that $A^r = 0$, the zero matrix, for some integer $r \geq 1$. Show that $I_n - A$ is invertible. [*Hint*: Compute the product

$$(I_n - A)(I_n + A + A^2 + \cdots + A^{r-1})$$

using the rules of matrix algebra.]

11. List the elements in $GL_2(F_2)$ where F_2 is the finite field of two elements, 0 and 1. (Give thanks that we used the smallest possible field!)

12. Explain succinctly why the solution space to a homogeneous system of m linear equations in n unknowns defined over a field k is a subspace of k^n. (*Hint*: Express the system in matrix form and relate it to a linear transformation from k^n to k^m.)

13. Express the following linear system as a single matrix equation and as a single vector equation, as shown in Section 5.2:

$$2x_1 - 4x_2 = 7$$
$$5x_1 + 9x_2 = 4$$

14. Without explicitly solving, show that the system above has a unique solution. (*Hint*: See Theorem 5-6.)

15. Without explicitly solving, show that the system

$$2x_1 + x_2 = y_1$$
$$-x_1 + 2x_2 = y_2$$

has a unique solution for all $y_1, y_2 \in R$.

16. Find the rank of the matrix

$$A = \begin{pmatrix} 1 & 2 & 1 \\ 2 & 0 & 2 \\ 0 & 0 & 1 \end{pmatrix}$$

17. Let A be a 4×7 matrix (over any field) with at least one nonzero entry. What are the possible values for the rank of A?

18. Using Gauss-Jordan elimination, solve the system given in Exercise 13.

19. Suppose that we are solving a 2×3 homogeneous linear system $Ax = 0$ by Gauss-Jordan elimination and reach the following augmented matrix in reduced row-echelon form:

$$\left(\begin{array}{ccc|c} 1 & 0 & -5 & 0 \\ 0 & 1 & 4 & 0 \end{array} \right)$$

What is the general solution to the original system? What is the dimension of the solution space?

20. Suppose that we are solving a 3×4 linear system $Ax = y$ by Gauss-Jordan elimination and reach the following augmented matrix in reduced row-echelon form:

$$\left(\begin{array}{cccc|c} 1 & 2 & 0 & 1 & 5 \\ 0 & 0 & 1 & 4 & 2 \\ 0 & 0 & 0 & 0 & 4 \end{array} \right)$$

What can one say about solutions to the original system?

21. Find all solutions to the following system by Gauss-Jordan elimination:

$$x_1 + x_2 - x_3 = 0$$
$$x_1 + 2x_2 + 4x_3 = 0$$

What is the dimension of the solution space?

22. Does every linear system for which there are more variables than equations have a solution? If not, what additional condition is needed?

23. Summarize in your own words why reduced row-echelon form is an effective device for solving linear systems of equations.

24. Factor the following matrix into the product of a lower triangular and an upper triangular matrix:

$$A = \begin{pmatrix} 2 & 5 \\ 1 & 2 \end{pmatrix}$$

[Hint: Try something of the form

$$A = \begin{pmatrix} 1 & 0 \\ x & 1 \end{pmatrix} \begin{pmatrix} y & z \\ 0 & w \end{pmatrix}$$

where x, y, z, and w are to be determined.]

25. Given the matrix factorization

$$\begin{pmatrix} 1 & 1 & 2 \\ 2 & 4 & 8 \\ 1 & 5 & 11 \end{pmatrix} = \begin{pmatrix} 1 & 0 & 0 \\ 2 & 1 & 0 \\ 1 & 2 & 1 \end{pmatrix} \begin{pmatrix} 1 & 1 & 2 \\ 0 & 2 & 4 \\ 0 & 0 & 1 \end{pmatrix}$$

solve the linear system

$$\begin{aligned} x_1 + x_2 + 2x_3 &= 11 \\ 2x_1 + 4x_2 + 8x_3 &= 30 \\ x_1 + 5x_2 + 11x_3 &= 28 \end{aligned}$$

by the method of LU decomposition.

26. Summarize in your own words why LU decomposition is an effective device for solving linear systems of equations.

27. Use the techniques of Section 5.4 to derive the general formula given in Exercise 7 above for the inversion of 2×2 matrices. (Be careful not to divide by zero; you may need to look at more than one case.)

28. Using the techniques of Section 5.4, invert the following carefully contrived matrix:

$$A = \begin{pmatrix} 1 & 2 & 2 \\ 2 & 4 & 3 \\ 3 & 5 & 4 \end{pmatrix}$$

This should go rather smoothly.

29. To the sound of the rain and the chamber music of Claude Debussy, the author reaches for his calculator, an old but serviceable *hp*-11C. Punching

the random number key nine times and recording the first digit to the right of the decimal point, he produces the following matrix:

$$A = \begin{pmatrix} 9 & 2 & 0 \\ 2 & 4 & 3 \\ 8 & 6 & 1 \end{pmatrix}$$

He ponders; he frowns; he leaves it to the student to find A^{-1}.

6
Representation of Linear Transformations

We saw in the previous chapter that a matrix in $\mathrm{Mat}_{m \times n}(k)$ acts by left multiplication as a linear transformation from k^n to k^m. In this chapter we shall see that in a strong sense every linear transformation of finite-dimensional vector spaces over k may be thus realized. (We say that the associated matrix *represents* the transformation.) In passing, we introduce the notion of a *k-algebra*, a rich structure that is a hybrid of both vector space and ring. We show that the set of linear transformations from an n-dimensional vector space to itself is in fact isomorphic as a k-algebra to the familiar matrix algebra $M_n(k)$.

The representation of a linear transformation almost always depends upon the choice of bases, and according to the transformation in question, some bases will serve us better than others both computationally and conceptually. (Chapters 9 and 10 are largely concerned with this theme.) Hence in Section 6.5 we also analyze just how the representation varies with this choice.

The material is difficult, but well worth the effort. It completes the bridge between the structural theory of vector spaces and the computational theory of matrices—a bridge that we will cross many times in later chapters.

6.1 The Space of Linear Transformations

Let V and W be vector spaces over k. Then $\mathrm{Hom}(V,W)$ denotes the set of all linear transformations from V to W. If $f \in \mathrm{Hom}(V,W)$ and $\lambda \in k$, we can define a function $\lambda f : V \to W$ by the formula

$$(\lambda f)(v) = \lambda \cdot f(v) \quad \forall v \in V$$

It is trivial to show that λf is likewise linear. Similarly, if also $g \in \mathrm{Hom}(V,W)$, we can define a function $f + g$ from V to W by

$$(f + g)(v) = f(v) + g(v) \quad \forall v \in V$$

and one shows easily that $f + g$ is linear. Thus $\mathrm{Hom}(V,W)$ is equipped with both addition and scalar multiplication.

6-1 PROPOSITION. Hom(V,W) *is a vector space over k with respect to the operations defined above.*

PROOF. We leave most of the details to the reader. Note that the zero in Hom(V,W) is the zero map, the constant function that sends everything in V to the zero vector of W. The additive inverse of $f \in$ Hom(V,W) is $(-1)f$. All of the requisite algebraic laws are inherited from the ambient space W. ☐

REMARK. This argument shows, moreover, that the set of functions from *any* nonempty set into W forms a vector space over k with respect to the operations given above. Thus neither the vector space structure of V nor the linearity of the elements of Hom(V,W) plays any role in this first result.

There is considerably more structure here than we have so far revealed; the precise statement is somewhat technical, but not deep.

6-2 PROPOSITION. *Let U, V, and W be vector spaces over k. Then*

(i) $g \circ (f_1 + f_2) = g \circ f_1 + g \circ f_2 \quad \forall g \in \text{Hom}(V,W), f_1, f_2 \in \text{Hom}(U,V)$

(ii) $(g_1 + g_2) \circ f = g_1 \circ f + g_2 \circ f \quad \forall g_1, g_2 \in \text{Hom}(V,W), f \in \text{Hom}(U,V)$

(iii) $(\lambda g) \circ f = g \circ (\lambda f) = \lambda(g \circ f) \quad \forall g \in \text{Hom}(V,W), f \in \text{Hom}(U,V), \lambda \in k$

PROOF. We shall only prove (i); the others are similar. Let $u \in U$. Then

$$
\begin{aligned}
g \circ (f_1 + f_2)(u) &= g((f_1 + f_2)(u)) \\
&= g(f_1(u) + f_2(u)) \\
&= g(f_1(u)) + g(f_2(u)) \\
&= g \circ f_1(u) + g \circ f_2(u)
\end{aligned}
$$

The calculation shows that both functions have the same effect on arbitrary elements of U. They are therefore identical. ☐

6-3 COROLLARY. *For any vector space V over a field k, Hom(V,V) is both a vector space over k (as described above) and a ring with unity (with respect to addition and composition of functions). Moreover, the vector space and ring structures are related by the following law:*

$$(\lambda g) \circ f = g \circ (\lambda f) = \lambda(g \circ f) \quad \forall f, g \in \text{Hom}(V,V), \lambda \in k$$

PROOF. This follows directly from the previous proposition in the special case that $U=V=W$. The unity in Hom(V,V) is the identity map. ☐

A structure that is both a ring and a vector space over k, with the additional property that the scalar multiplication commutes with the ring multiplication in the sense of the preceding corollary, is called a *k-algebra*. We have therefore shown that $\text{Hom}(V,V)$ is a k-algebra for all vector spaces V over k. Note that according to Propositions 5-1 and 5-2, the matrix ring $M_n(k)$ is also a k-algebra for all positive n. (And surely the discovery of two k-algebras in the same paragraph must be fraught with latent but portentous consequences.)

The final result of this section is crucial to the sequel.

6-4 PROPOSITION. *Let f and g lie in* $\text{Hom}(V,W)$. *If f and g take identical values on the elements of a spanning set for V, then they are identical on all of V; that is, f=g.*

PROOF. Let S be a spanning set for V such that f and g agree on S. Let v be any element of V. Then we may write v as a linear combination of elements in S:

$$v = \sum_{j=1}^{n} \lambda_j v_j$$

Since f and g are linear, we have

$$f(v) = f\left(\sum_{j=1}^{n} \lambda_j v_j\right) = \sum_{j=1}^{n} \lambda_j f(v_j) = \sum_{j=1}^{n} \lambda_j g(v_j) = g\left(\sum_{j=1}^{n} \lambda_j v_j\right) = g(v)$$

The middle step is justified by the assumption that f and g take the same values on S. This completes the proof.   ☐

6.2 The Representation of $\text{Hom}(k^n,k^m)$

In this section we deal exclusively with the vector spaces k^n and k^m, whose elements we regard as column vectors. We saw previously that with every $m \times n$ matrix A we could associate the linear transformation $T_A \in \text{Hom}(k^n,k^m)$ defined by $T_A(x) = Ax$. We now reverse the association to show every linear transformation in $\text{Hom}(k^n,k^m)$ is realizable as left multiplication by the appropriate matrix.

DEFINITION. Let T be an arbitrary linear transformation from k^n to k^m. Define $M(T) \in \text{Mat}_{m \times n}(k)$ to be the matrix whose jth column is the vector $T(e_j)$. We call $M(T)$ the *matrix of T with respect to the canonical bases for* k^n *and* k^m.

EXAMPLE. Consider the linear transformation $T: \mathbf{R}^3 \to \mathbf{R}^2$ defined by the following rule of assignment:

$$\begin{pmatrix} x_1 \\ x_2 \\ x_3 \end{pmatrix} \overset{T}{\mapsto} \begin{pmatrix} 2x_1 + x_3 \\ x_2 - x_3 \end{pmatrix}$$

Then

$$T(e_1) = \begin{pmatrix} 2 \\ 0 \end{pmatrix}, \quad T(e_2) = \begin{pmatrix} 0 \\ 1 \end{pmatrix}, \quad T(e_3) = \begin{pmatrix} 1 \\ -1 \end{pmatrix}$$

Whence

$$M(T) = \begin{pmatrix} 2 & 0 & 1 \\ 0 & 1 & -1 \end{pmatrix}$$

In anticipation of the general result to follow, we note a special relationship between T and $M(T)$: for all $x \in \mathbf{R}^3$, $T(x)$ has the same value as the matrix product $M(T)x$, as seen by the following simple calculation:

$$\begin{pmatrix} 2 & 0 & 1 \\ 0 & 1 & -1 \end{pmatrix} \begin{pmatrix} x_1 \\ x_2 \\ x_3 \end{pmatrix} = \begin{pmatrix} 2x_1 + x_3 \\ x_2 - x_3 \end{pmatrix}$$

Hence to apply the transformation T, we might just as well multiply by $M(T)$. A more elegant way of saying the same thing, using our earlier association of a linear transformation with a matrix, is this: $T = T_{M(T)}$. We now formalize and extend these ideas.

6-5 PROPOSITION. *The mappings*

$$\mathrm{Mat}_{m \times n}(k) \rightarrow \mathrm{Hom}(k^n, k^m) \qquad \mathrm{Hom}(k^n, k^m) \rightarrow \mathrm{Mat}_{m \times n}(k)$$
$$A \mapsto T_A \qquad\qquad\qquad T \mapsto M(T)$$

are mutually inverse isomorphisms of vector spaces. In particular, given any linear transformation $T \in \mathrm{Hom}(k^n, k^m)$, $M(T)$ is the unique matrix such that

$$T(x) = M(T)x \quad \forall x \in k^n$$

The result says, in effect, that every linear transformation $T : k^n \rightarrow k^m$ is uniquely *represented* by $M(T)$ in the sense that T has the same effect as left multiplication by $M(T)$. This association is so compelling that often we fail to distinguish between T and $M(T)$.

PROOF. We shall show that the given maps are inverse to each other, leaving linearity to the student. (One direction suffices. Why?) For this we ask two questions:

(i) What is $M(T_A)$ for any given $m \times n$ matrix A? By definition, the jth column of $M(T_A)$ is precisely $T_A(e_j) = Ae_j = A^j$, the jth column of A. (See Eq. 5.2 in the previous chapter.) But if $M(T_A)$ and A have exactly the same columns, they are the same matrix. Hence $M(T_A) = A$.

(ii) What is $T_{M(T)}$ for any given linear transformation T? By definition $T_{M(T)}(e_j)$ is the matrix product $M(T)e_j$, which is the jth column of $M(T)$. But the jth column of $M(T)$ is precisely $T(e_j)$, and this means that $T_{M(T)}$ and T have the same effect on the canonical basis. By Proposition 6-4, they must be the same function. Hence $T_{M(T)} = T$.

Parts (i) and (ii) taken together indeed show that the given maps are mutually inverse, and (ii) says precisely that left multiplication by $M(T)$ is equivalent to T, as claimed. Uniqueness follows from the injectivity of the map $A \mapsto T_A$: distinct matrices induce distinct linear maps, so no other matrix but $M(T)$ can represent T.   □

Since isomorphisms preserve dimension (a basis maps to a basis) and the dimension of $\mathrm{Mat}_{m \times n}(k)$ is mn, this proposition yields the following corollary:

6-6 COROLLARY. *The dimension of* $\mathrm{Hom}(k^n, k^m)$ *is* mn.   □

Again, there is more to be said with regard to composition. Let $A \in \mathrm{Mat}_{m \times n}(k)$ and $B \in \mathrm{Mat}_{n \times p}(k)$. Then both T_{AB} and $T_A \circ T_B$ lie in $\mathrm{Hom}(k^p, k^m)$. How do these maps compare? Let us compute their effect on the canonical basis vectors. First,

$$T_{AB}(e_j) = (AB)e_j = (AB)^j$$

the jth column of AB. Second,

$$T_A \circ T_B(e_j) = T_A(T_B(e_j)) = T_A(Be_j) = T_A(B^j) = A \cdot B^j$$

which is A times the jth column of B. But by Eq. 5.1, these two expressions are equal. Hence T_{AB} and $T_A \circ T_B$ agree on a spanning set for k^p and therefore agree everywhere! We state this fundamental property as a proposition. It has two striking corollaries.

6-7 PROPOSITION. $T_{AB} = T_A \circ T_B$ whenever the indicated matrix product is defined.   □

CRITICAL REMARK. *Matrix multiplication is defined the way it is precisely so that this result holds; that is, precisely to reflect the composition of linear transformations.*

We immediately exploit this proposition to give a truly elegant, non-computational proof of the most fundamental property of matrix arithmetic.

6-8 COROLLARY. *Matrix multiplication is associative.*

PROOF. Suppose that the products $A(BC)$ and $(AB)C$ are defined. Then according to the previous proposition,

$$T_{A(BC)} = T_A \circ T_{BC} = T_A \circ (T_B \circ T_C)$$

and

$$T_{(AB)C} = T_{AB} \circ T_C = (T_A \circ T_B) \circ T_C$$

But these maps are equal by the associativity of composition. This shows that

$$T_{A(BC)} = T_{(AB)C}$$

Since according to Proposition 6-5 the association $A \mapsto T_A$ is bijective, we must have that $A(BC) = (AB)C$ as claimed. Hence the associativity of matrix multiplication is just an aspect of the associativity of composition of functions.   □

Perhaps not surprisingly, a tight relationship between matrix multiplication and composition of linear transformations also holds in the direction opposite to that expressed in Proposition 6-7.

6-9 COROLLARY. *Let $T: k^p \to k^n$ and $T': k^n \to k^m$ be linear transformations. Then*

$$M(T' \circ T) = M(T')M(T)$$

Thus composition of linear transformations corresponds to matrix multiplication.

PROOF. The key to the proof is the associativity of matrix multiplication. Given any $x \in k^p$, since $M(T')$ and $M(T)$ represent T' and T, respectively, we have

$$T' \circ T(x) = T'(T(x))$$
$$= T'(M(T)x)$$
$$= M(T')(M(T)x)$$
$$= (M(T')M(T))x$$

It follows that the product $M(T')M(T)$ represents $T' \circ T$. By the uniqueness of such a representation, it follows that $M(T' \circ T) = M(T')M(T)$. $\square$

If M and N are k-algebras, then a vector space isomorphism $\varphi : M \to N$ is called an *algebra isomorphism* (or an *isomorphism of k-algebras*) if it satisfies the additional condition

$$\varphi(ab) = \varphi(a)\varphi(b) \quad \forall a,b \in M$$

With this definition, the results of the current section culminate in a fundamental theorem on the representation of linear transformations from the vector space k^n to itself.

6-10 THEOREM. *The pair of mappings*

$$M_n(k) \to \text{Hom}(k^n, k^n) \qquad \text{Hom}(k^n, k^n) \to M_n(k)$$
$$A \mapsto T_A \qquad\qquad\qquad T \mapsto M(T)$$

are mutually inverse isomorphisms of k-algebras.

PROOF. This statement summarizes Propositions 6-5 and 6-7 in the special case $m=n=p$. Proposition 6-5 says that the given maps are inverse isomorphisms of vector spaces. Proposition 6-7 says that, moreover, they are isomorphisms of k-algebras: matrix multiplication in $M_n(k)$ corresponds to composition of functions in $\text{Hom}(k^n, k^n)$. $\square$

EXAMPLE. Let $V = \mathbf{R}^2$ and let T_θ be the map which rotates points in the plane counterclockwise around the origin by the angle θ. One can see geometrically that this map is linear. (For instance, in light of the parallelogram law for vector addition, it is easy to analyze the effect of rotation on the sum of two vectors.) We compute its matrix relative to the canonical basis. Rotating both canonical basis vectors by θ (draw a picture!), we have at once that

$$T_\theta(e_1) = (\cos\theta, \sin\theta)$$
$$T_\theta(e_2) = (-\sin\theta, \cos\theta)$$

Therefore the matrix of the transformation is

$$M(T_\theta) = \begin{pmatrix} \cos\theta & -\sin\theta \\ \sin\theta & \cos\theta \end{pmatrix}$$

We can use this to prove two elementary trigonometric identities. Let ψ be a second angle. Then by the nature of rotation, $T_{\theta+\psi} = T_\theta \circ T_\psi$ whence

$$M(T_{\theta+\psi}) = M(T_\theta)M(T_\psi)$$

This amounts to the matrix equation

$$\begin{pmatrix} \cos(\theta+\psi) & -\sin(\theta+\psi) \\ \sin(\theta+\psi) & \cos(\theta+\psi) \end{pmatrix} = \begin{pmatrix} \cos\theta & -\sin\theta \\ \sin\theta & \cos\theta \end{pmatrix}\begin{pmatrix} \cos\psi & -\sin\psi \\ \sin\psi & \cos\psi \end{pmatrix}$$

Multiplying out the right-hand product and comparing entries yields the well-known (but usually forgotten) identities:

$$\cos(\theta+\psi) = \cos\theta\cos\psi - \sin\theta\sin\psi$$
$$\sin(\theta+\psi) = \sin\theta\cos\psi + \cos\theta\sin\psi$$

(This is the best way to remember these identities!)

6.3 The Representation of Hom(V,V')

Throughout this section, all vector spaces are finite dimensional.

Let $T : V \to V'$ be a linear transformation of vector spaces over k. Again we would like to represent this map by a matrix, but as we have seen in Section 6.2, this requires (and will depend upon) a choice of coordinate systems for both V and V'.

Let $\dim(V) = n$ and let $B = \{v_1,\ldots,v_n\}$ be a basis for V. (We should speak more precisely of an *ordered basis* in this context, since the ordering of the basis vectors in B affects the subsequent representation.) We have seen that the co-ordinate map $\gamma_B : V \to k^n$ is the isomorphism defined by

$$\gamma_B\left(\sum_{j=1}^n \lambda_j v_j\right) = (\lambda_1,\ldots,\lambda_n)$$

Recall that this makes sense because every vector in V can be expressed uniquely as a linear combination of the basis elements. (Note that in the sequel we shall usually regard these coordinate vectors as columns, even though we write them as rows for typographical efficiency.) In the same way, if further-more $\dim(V')=m$ and V' has basis $B' = \{v'_1,\ldots,v'_m\}$, we have the coordinate map $\gamma_{B'}: V' \to k^m$. Now since an isomorphism is invertible, we can construct the fol-lowing commutative diagram:

$$
\begin{array}{ccc}
& T & \\
V & \longrightarrow & V' \\
\gamma_B \downarrow & & \downarrow \gamma_{B'} \\
k^n & \longrightarrow & k^m \\
& \gamma_{B'} \circ T \circ \gamma_B^{-1} &
\end{array}
$$

The bottom map is unique in the sense that it is the only way to complete the square. Moreover, being a linear transformation from k^n to k^m, it may be repre-sented by a matrix. This leads us to a fundamental definition.

DEFINITION. Assuming the notation of the previous paragraphs, the *matrix of T with respect to the bases B and B'* is the matrix of $\gamma_{B'} \circ T \circ \gamma_B^{-1}$ in the sense of the previous section. We denote this matrix $M_{B,B'}(T)$.

Referring again to the diagram above, we see that $M_{B,B'}(T)$ is the unique matrix such that

$$\gamma_{B'}(T(v)) = M_{B,B'}(T)\gamma_B(v)$$

for all $v \in V$. In this sense $M_{B,B'}(T)$ *represents* T, at the cost of introducing the coordinate maps. With this in mind, we might recast our previous commutative diagram as follows:

$$
\begin{array}{ccc}
& T & \\
V & \longrightarrow & V' \\
\gamma_B \downarrow & & \downarrow \gamma_{B'} \\
k^n & \longrightarrow & k^m \\
& M_{B,B'}(T) &
\end{array}
$$

The label on the bottom arrow indicates that the corresponding map is precisely matrix multiplication by $M_{B,B'}(T)$.

The preceding definition, while somewhat technical, will reveal its elegance in subsequent proofs. The actual practice of constructing the matrix of a linear transformation is straightforward, as we shall soon see. But before moving along to the examples, we establish some formal properties of this construction similar to those proved in Section 6.2.

6-11 THEOREM. *Let V and V' be finite-dimensional vector spaces over the field k, with respective bases B and B' and dimensions n and m. Then the map*

$$\text{Hom}(V,V') \to \text{Mat}_{m \times n}(k)$$
$$T \mapsto M_{B,B'}(T)$$

is an isomorphism of vector spaces.

PROOF. Note that the proof has a slightly different flavor from that of Proposition 6-5, since the inverse map in this case is a bit harder to work with. Leaving linearity as an exercise, we begin by noting that $M_{B,B'}$ is at least injective since it has trivial kernel: the only linear transformation represented by the zero matrix is the zero transformation. It only remains to show that $M_{B,B'}$ is surjective, and this amounts to showing that every matrix $A \in \text{Mat}_{m \times n}(k)$ represents some linear transformation T from V to V' relative to the bases B and B'. But referring to the previous commutative diagram, which essentially defines the matrix of a transformation, it is clear that A represents the linear transformation

$$V \to V'$$
$$v \mapsto \gamma_{B'}^{-1} \circ T_A \circ \gamma_B(v)$$

[If this is not clear, apply the definition of $M_{B,B'}$ to see that the matrix of this map reduces to $M(T_A)=A$.] Hence $M_{B,B'}$ is surjective, as required.   ☐

6-12 COROLLARY. *Let V and V' have dimensions n and m, respectively. Then the dimension of* Hom(*V*,*V'*) *is mn.*   ☐

With regard to composition, we have an amiable generalization of Proposition 6-7.

6-13 PROPOSITION. *Let there be given vector spaces V, V', and V'' over a common field k, with bases B, B', and B'', respectively, and linear transformations T:V→V' and T':V'→V''. Then*

$$M_{B,B''}(T' \circ T) = M_{B',B''}(T')M_{B,B'}(T)$$

PROOF. The proof is a simple calculation based on the corresponding fact for the composition of linear transformations from k^n to k^m:

$$M_{B,B''}(T' \circ T) = M(\gamma_{B''} \circ (T' \circ T) \circ \gamma_B^{-1})$$
$$= M(\gamma_{B''} \circ T' \circ \gamma_{B'}^{-1} \circ \gamma_{B'} \circ T \circ \gamma_B^{-1})$$
$$= M(\gamma_{B''} \circ T' \circ \gamma_{B'}^{-1}) M(\gamma_{B'} \circ T \circ \gamma_B^{-1})$$
$$= M_{B',B''}(T') M_{B,B'}(T) \qquad \square$$

REMARK. Using the commutative diagram of Figure 6.1 below, we may give an appealing, alternative proof of this proposition. The diagram shows at once that

$$\gamma_{B''} \circ T' \circ T(v) = M_{B',B''}(T') M_{B,B'}(T) \gamma_B(v)$$

for all $v \in V$. But then the matrix product on the right must be $M_{B,B''}(T' \circ T)$ since this is the unique matrix representing $T' \circ T$ with respect to the bases B and B''.

$$
\begin{array}{ccccc}
 & T & & T' & \\
V & \longrightarrow & V' & \longrightarrow & V'' \\
\gamma_B \downarrow & & \downarrow \gamma_{B'} & & \downarrow \gamma_{B''} \\
k^p & \longrightarrow & k^n & \longrightarrow & k^m \\
 & M_{B,B'}(T) & & M_{B',B''}(T') &
\end{array}
$$

Figure 6.1. The diagram suggests an alternative proof of Proposition 6-13.

We pass now to the important special case that $V = V'$ with a single given basis B. If $T \in \mathrm{Hom}(V,V)$, we abbreviate $M_{B,B}(T)$ to $M_B(T)$, which is then the matrix of $\gamma_B \circ T \circ \gamma_B^{-1}$. This is called simply *the matrix of T relative to B*. The following theorem summarizes all of our results so far.

6-14 THEOREM. *Let V be an n-dimensional vector space over k, and let B be any basis for V. Then the mapping*

$$\mathrm{Hom}(V,V) \to M_n(k)$$
$$T \quad \mapsto M_B(T)$$

is an isomorphism of k-algebras. $\qquad \square$

In preparation for the following examples, we note that the general procedure for constructing the matrix of a linear transformation T relative to a pair of bases B and B' is quite simple and falls right out of the definition. We state it as follows:

RULE. *The jth column of $M_{B,B'}(T)$ is the coordinate vector relative to B' of $T(v_j)$, where v_j is the jth element of the basis B.*

To verify this, observe that by the definition of $M_{B,B'}(T)$,

$$M_{B,B'}(T)\gamma_B(v_j) = \gamma_{B'}(T(v_j))$$

But $\gamma_B(v_j)$ is precisely e_j, so

$$M_{B,B'}(T)\gamma_B(v_j) = M_{B,B'}(T)e_j$$

which we recognize as the jth column of $M_{B,B'}(T)$. Hence these last two equations show that the jth column of $M_{B,B}(T)$ is precisely the coordinate vector of $T(v_j)$ relative to B', as claimed.

We now illustrate the theory for linear transformations of a finite-dimensional vector space to itself.

EXAMPLES

(1) Let V be the real function space defined by the span of the functions $\sin x$ and $\cos x$, which therefore constitute a basis for V. Let D be the differentiation map on V. We compute the matrix of D with respect to the given basis. The plan is to evaluate D on members of the basis, to find the resulting coordinate vectors, and to install these coordinate vectors as the columns of the matrix of D. We may combine the first two steps as follows:

$$D(\sin x) = \quad 0 \cdot \sin x + 1 \cdot \cos x$$
$$D(\cos x) = -1 \cdot \sin x + 0 \cdot \cos x$$

Hence according to the rule above, the matrix of D is

$$A = \begin{pmatrix} 0 & -1 \\ 1 & 0 \end{pmatrix}$$

Let us check the meaning of this. Consider the function

$$f(x) = 2\sin x + 5\cos x$$

In principle, we should be able to compute the derivative of f using the matrix A, as follows. First, find the coordinate vector of f relative to our basis. This is $(2,5)$—but be prepared to think of it as a column. Second, multiply the coordinate vector by A:

$$\begin{pmatrix} 0 & -1 \\ 1 & 0 \end{pmatrix}\begin{pmatrix} 2 \\ 5 \end{pmatrix} = \begin{pmatrix} -5 \\ 2 \end{pmatrix}$$

Finally, interpret the result of this calculation as a set of coordinates in V. What function does it represent? The vector components take their respective places as coefficients of the basis elements. Thus

$$f'(x) = -5\sin x + 2\cos x$$

which is correct. Admittedly, this is the long way around to differentiate f, but it does illustrate the point nicely.

(2) Continuing the example above, what is the matrix of D^2, the second derivative operator? Since $D^2 = D \circ D$, we can proceed in either of two ways: calculate directly as above or, applying Proposition 6-13, square the matrix A computed previously. Choosing the latter, we find that

$$A^2 = \begin{pmatrix} -1 & 0 \\ 0 & -1 \end{pmatrix}$$

The student should also make the direct calculation to verify that the same result is obtained. Along similar lines, D^4, the fourth-derivative operator, evidently yields the identity map on V. Indeed one can verify easily by matrix arithmetic that $A^4 = I_2$.

(3) Let V be the real vector space of polynomials in x of degree less than or equal to 4, with basis 1, x, x^2, x^3, x^4. One shows easily that the matrix of the differentiation operator D on V is

$$A = \begin{pmatrix} 0 & 1 & 0 & 0 & 0 \\ 0 & 0 & 2 & 0 & 0 \\ 0 & 0 & 0 & 3 & 0 \\ 0 & 0 & 0 & 0 & 4 \\ 0 & 0 & 0 & 0 & 0 \end{pmatrix}$$

Since the fifth-derivative operator is the zero map on V, it follows that A^5 is the 5×5 zero matrix. The diligent student should verify this.

6.4 The Dual Space

This section looks at a special case of $\mathrm{Hom}(V,V')$, namely when V' is k, the ground field itself. The analysis seems entirely formal, but nonetheless it explains one of the more surprising properties of matrices (Corollary 6-20) and deepens our understanding of linear systems (Theorem 6-21). First we shall need a fundamental result which explains some hitherto mysterious terminology introduced in Chapter 4.

6-15 THEOREM. *Let V be a vector space over k with basis B. Then given any vector space W over k and any function $g:B \to W$, there is a unique linear transformation $\bar{g}:V \to W$ which extends g in the sense that $\bar{g}(v) = g(v)$ for all $v \in B$; that is, the following diagram is commutative:*

$$
\begin{array}{ccc}
B & \xrightarrow{\ i\ } & V \\
{\scriptstyle g}\downarrow & \swarrow{\scriptstyle \bar{g}} & \\
W & &
\end{array}
$$

The top map $i : B \to V$ is the function that sends every element in B to itself—much like the identity function—but here the codomain is V. This is called the *inclusion map*.

PROOF. Given any vector $v \in V$, it can be written uniquely as a linear combination of basis elements. Say

$$
v = \sum_{j=1}^{n} \lambda_j v_j \quad (v_1, \ldots, v_n \in B)
$$

Then define $\bar{g}$ as follows:

$$
\bar{g}(v) = \sum_{j=1}^{n} \lambda_j g(v_j)
$$

This is called *extension by linearity*. We leave it to the reader to show that this map is indeed linear and that it extends g in the sense above. That it is unique is clear: a linear function is completely determined by its effect on a spanning set (in this case B), and hence no other extension is possible.   □

Notice how this result depends upon the linear independence of a basis. For example, given the vectors v and $-v$, no such assertion is possible: there are many set maps g from $\{v, -v\}$ into an arbitrary vector space W, but there is no chance of extending g to a linear map on V unless $g(-v) = -g(v)$. The vectors are bound by a dependence relation. In contrast, the elements of a basis are completely unbound—any mapping of them into W can be extended to a linear transformation on all of V. Hence one says that a vector space is *free on its basis*, and since every vector space has a basis, one says that *vector spaces are free*. This explains the terminology of Chapter 4.

We shall soon put this theorem to work in connection with the following construction.

DEFINITION. Let V be a vector space over k. Then $\text{Hom}(V,k)$, the set of all vector space homomorphisms from V into k, is called the *dual space of V* and denoted V^*.

Both notations $\text{Hom}(V,k)$ and V^* are suggestive and useful. We shall pass from one to the other as appropriate.

The definition above associates with a vector space V over k its dual space V^*, but this is only half of the story. Let $T: V \to W$ be a linear transformation. Then given any element $f \in \text{Hom}(W,k)$, the composed map $f \circ T$ is a homomorphism from V to k. Thus we have defined a map, which is itself a linear transformation,

$$T^*: W^* \to V^*$$
$$f \mapsto f \circ T$$

T^* is called the *transpose map* of T. Transposition has the following elementary properties, which we leave as exercises:

(i) For any vector space V over k, $(1_V)^* = 1_{V^*}$; that is, the transpose of the identity map on V is the identity map on V^*.

(ii) Let T_1 and T_2 be linear transformations of vector spaces over k such that the composition $T_1 \circ T_2$ is defined. Then

$$(T_1 \circ T_2)^* = T_2^* \circ T_1^*$$

The reversal of the order of composition is known as *contravariance*, and $\text{Hom}(V,-)$ is in fact our first example of a *contravariant functor*, although we shall say nothing further about this for the present (cf. Supplementary Topics).

The transpose operator has certain critical formal properties, some of which we begin to explore in the following result.

6-16 PROPOSITION. *Let $T: V \to W$ be a linear transformation of vector spaces over k. Then*

(i) *if T is surjective, then T^* is injective;*

(ii) *if T is injective, then T^* is surjective.*

PROOF. (i) Let f and g lie in $W^* = \text{Hom}(W,k)$ and suppose that $T^*(f) = T^*(g)$. This means that $f \circ T = g \circ T$. Since T is assumed surjective, given any $w \in W$, there exists $v \in V$ such that $T(v)=w$. Then

$$f(w) = f(T(v)) = f \circ T(v) = g \circ T(v) = g(T(v)) = g(w)$$

and hence $f=g$ as required. Notice that this argument requires nothing but set theory.

(ii) Let $f \in \text{Hom}(V,k)$. Then we must show that $f=T^*(g)$ for some $g \in \text{Hom}(W,k)$. Since $T^*(g)=g \circ T$, this means that we must demonstrate the existence of a linear transformation g which makes the following diagram commutative:

Notice the similarity of this picture to that of the first theorem in this section. Given any basis B for V, we know that the injective map T sends the linearly independent set B to a linearly independent set $T(B) \subseteq W$, which, while not necessarily a basis for W, may be extended to a basis for W. Call this basis B'. Define a set map from B' to k according to the following dichotomy:

if $w=T(v)$ for some $v \in B$, then send w to $f(v)$; otherwise, send w to 0

In other words, send the elements in B' that came from B via T to the corresponding value of the original function f and send the rest to 0 (or anywhere else for that matter). By Proposition 6-15, this set map from $B' \to k$ extends uniquely to a linear transformation $g : W \to k$. By construction, f and $g \circ T$ agree on the basis B for V and hence agree on all of B. This completes the proof. ☐

This result leads immediately to a major theorem.

6-17 THEOREM. *Let $T:V \to W$ be a linear transformation of finite-dimensional vector spaces over k. Then*

$$\text{rk}(T) = \text{rk}(T^*)$$

That is, the dimension of the image of T is equal to the dimension of the image of T^*.

We shall use this shortly to give a most elegant proof that a matrix and its transpose have the same rank.

PROOF. First, a special case. Suppose that T is surjective. Then $\mathrm{Im}(T)=W$, and $\mathrm{rk}(T)=\dim(W)$. By Proposition 6-16, the transpose map $T^*:W^*\to V^*$ is injective and thus has trivial kernel. From the Rank-Nullity Theorem we may therefore infer that $\mathrm{rk}(T^*)=\dim(W^*)$. But we shall see shortly (Proposition 6-18) that in fact $\dim(W^*)=\dim(W)$, whence $\mathrm{rk}(T)=\dim(W)=\dim(W^*)=\mathrm{rk}(T^*)$, completing the proof for this case.

Now let T be arbitrary and let $\mathrm{Im}(T)=W_1$, a subspace of W. We construct the following commutative diagram, which factors T into the composition of a surjective map followed by an injective map:

$$\begin{array}{ccc} & & W \\ & {}^{T}\nearrow & \uparrow i \\ V & \xrightarrow{\quad} & W_1 \\ & T_1 & \end{array}$$

The map T_1 is just the same rule of assignment as T but with codomain reassigned to W_1, the image of the original map T. The map i is just inclusion of W_1 into W. Note that T_1 is by construction surjective and i is evidently injective. Observe also that $\mathrm{rk}(T) = \mathrm{rk}(T_1)$ since in each case the image is W_1.

Now take the dual of each space in the diagram and the transpose of each corresponding map. By contravariance we obtain a second commutative diagram as follows:

$$\begin{array}{ccc} & & W^* \\ & {}^{T^*}\swarrow & \downarrow i^* \\ V^* & \xleftarrow{\quad} & W_1^* \\ & T_1^* & \end{array}$$

By the previous proposition, i^* is surjective. By set theory it follows that T^* and T_1^* have the same image and hence the same rank. Thus we have shown so far that

$$\mathrm{rk}(T) = \mathrm{rk}(T_1) \quad \text{and} \quad \mathrm{rk}(T^*) = \mathrm{rk}(T_1^*)$$

But T_1 is surjective and thus conforms to our special case. Therefore

$$\mathrm{rk}(T_1) = \mathrm{rk}(T_1^*)$$

and by transitivity of equality $\mathrm{rk}(T)=\mathrm{rk}(T^*)$, as claimed.   □

The Dual Basis

No doubt the student has wondered at the coincidence of language in speaking of the transpose of a matrix and the transpose of a linear map. We shall now explain this and at the same time prove a beautiful result.

Let V be a finite-dimensional vector space with basis $v_1,...,v_n$. In V^* we define the elements $v_1^*,...,v_n^*$ to be the unique linear maps from V to k whose effect on the given basis is

$$v_j^*(v_k) = \delta_{jk}$$

Thus v_j^* takes the value 1 on the basis vector v_j, zero on every other basis vector, and extends by linearity to all of V. Given any $f \in V^*$, one sees at once that

$$f = \sum_{j=1}^{n} f(v_j)v_j^*$$

since by construction both sides clearly agree on the basis $v_1,...,v_n$. Moreover, this representation of f is obviously unique. Hence every element of V^* can be expressed in exactly one way as a linear combination of the v_j^*, which therefore constitute a basis for V^*. This is called the *dual basis* to $v_1,...,v_n$.

6-18 PROPOSITION. *In the context above, the linear transformation from V to V^* defined by*

$$v_j \mapsto v_j^*$$

is an isomorphism of vector spaces. Hence $V \cong V^*$.

(Recall that this mapping of basis elements yields a unique linear transformation from V to V' via extension by linearity.)

PROOF. We can clearly define an inverse map by $v_j^* \mapsto v_j$; hence the given map is bijective.   ◻

One sometimes says that the given isomorphism is not *natural* because it depends upon a choice of basis. We shall see later an example of a natural isomorphism, although we shall not attempt a technical definition of this subtle and historically elusive term.

Resuming the general discussion, if $B_1 = \{v_1,...,v_n\}$ is a basis for the vector space V_1 and $B_2 = \{w_1,...,w_m\}$ is a basis for the vector space V_2, we let B_1^* and B_2^* denote, respectively, the corresponding dual bases for V_1^* and V_2^*. Now let $T:V_1 \rightarrow V_2$ be a linear transformation. The following result discloses the connection between the transpose operator for linear transformations and the transpose operator for matrices.

6-19 THEOREM. *Let $M(T)$ denote the matrix of T with respect to the bases B_1 and B_2 and let $M(T^*)$ denote the matrix of T^* with respect to the dual bases B_2^* and B_1^*. Then*

$$M(T^*) = {}^t M(T)$$

That is, the transpose of the matrix of T with respect to the given bases is the matrix of the transpose of T with respect to the dual bases.

Before proving this, we note an immediate corollary.

6-20 COROLLARY. *The rank of a matrix is equal to the rank of its transpose. Hence the dimension of the space spanned by the columns of a given matrix is equal to the dimension of the space spanned by its rows.*

PROOF OF COROLLARY. The rank of a matrix A is just the rank of the associated linear transformation T_A. We have seen that the rank of T_A is equal to the rank of T_A^*, which by the theorem to be proven is represented by ${}^t A$ with respect to the dual basis. Hence $\text{rk}(A)=\text{rk}(T_A)=\text{rk}(T_A^*)=\text{rk}({}^t A)$.   □

PROOF OF THEOREM. By definition, the jth column of the matrix of T^* with respect to the dual bases B_2^* and B_1^* is the coordinate vector of $T^*(w_j^*)$ with respect to $v_1^*,\dots,v_n^*$. This is not hard to compute if we keep in mind three elementary facts:

(i)   As we saw above, any linear map from $f: V \to k$ may be expressed in terms of the dual basis as

$$f = \sum_{j=1}^{n} f(v_j) v_j^*$$

Again, this is because both of these maps obviously agree on the given basis $v_1,\dots,v_n$.

(ii) If $M(T)=(a_{ij})$ with respect to the bases B_1 and B_2, then

$$T(v_j) = \sum_{i=1}^{m} a_{ij} w_i$$

since the jth column of $M(T)$ is by definition the set of coefficients needed to express $T(v_j)$ as a linear combination of the basis elements $w_1,\dots,w_m$.

(iii) By definition of the dual basis,

$$w_j^*(\sum_{i=1}^{m} \lambda_i w_i) = \lambda_j$$

since w_j^* annihilates all but the w_j component of the summation.

Using (i)-(iii) in succession (changing indices where necessary), the calculation of $T^*(w_j^*)$ is now entirely straightforward:

$$T^*(w_j^*) = w_j^* \circ T$$

$$= \sum_{i=1}^{n} [w_j^* \circ T](v_i) v_i^*$$

$$= \sum_{i=1}^{n} w_j^* (T(v_i)) v_i^*$$

$$= \sum_{i=1}^{n} w_j^* (\sum_{k=1}^{m} a_{ki} w_k) v_i^*$$

$$= \sum_{i=1}^{n} a_{ji} v_i^*$$

And now the startling conclusion: the jth column of $M(T^*)$ is the jth row of $M(T)$! Thus $M(T^*) = {}^t M(T)$, as claimed.   ☐

The corollary stated above allows us at once to extend impressively the already ponderous Theorem 5-6. (Even so, this is not yet the end of the story for this theorem. Still more awaits us in Chapter 8.)

6-21 THEOREM. *Let $A \in M_n(k)$. Then the following ten statements are equivalent:*

(i)   *The linear system $Ax=y$ has at least one solution for all $y \in k^n$.*

(ii) *The columns of A span k^n.*

(iii) *The rows of A span k^n.*

(iv) *The homogeneous linear system $Ax=0$ has only the trivial solution $x=0$.*

(v)   *The columns of A are linearly independent.*

(vi) *The rows of A are linearly independent.*

(vii) *The linear system $Ax=y$ has exactly one solution for all $y \in k^n$.*

(viii) *The columns of A constitute a basis for k^n.*

(ix) *The rows of A constitute a basis for k^n.*

(x) *A is invertible; i.e., $A \in GL_n(k)$.*

PROOF. In light of the equality of the rank of A with the rank of its transpose, the first nine statements follow at once from the general proposition that for n vectors in an n-dimensional space, the attributes of linear independence, spanning, and being a basis are equivalent. Moreover, we have seen in the earlier version of this theorem (5-6) that (x) implies (vii). Hence it only remains to show that a matrix satisfying (i)-(ix) is invertible.

We observed in Section 5.2 that (i) implies the existence of a right inverse B for A. We now produce a left inverse analogously and show that the two are equal. First note how left multiplication of A by a row vector x operates:

$$(x_1,\ldots,x_n)\begin{pmatrix} a_{11} & a_{12} & \cdots & a_{1n} \\ a_{21} & a_{22} & \cdots & a_{2n} \\ \vdots & \vdots & & \vdots \\ a_{n1} & a_{n2} & \cdots & a_{nn} \end{pmatrix} = x_1 A_1 + \cdots + x_n A_n$$

The result is that the coordinates of x appear on the right as coefficients in a linear combination of the rows of A. Thus one can interpret (iii) to mean that for every canonical basis vector e_j (considered here as a row vector), there exists a row vector C_j such that $C_j A = e_j$ ($j=1,\ldots,n$). Stacking these rows into a matrix C, we have by construction that $CA = I_n$. Finally note that $C=CAB=B$, showing that A is indeed invertible and accordingly lies in $GL_n(k)$.   □

The Dual of the Dual

We are left with one loose end. Given a finite-dimensional vector space V over k, we have constructed the dual space V^* and have shown that V and V^* are isomorphic, although the isomorphism does depend upon the choice of a basis. One wonders, what is V^{**}, the dual of the dual? It is of course isomorphic to V^* and hence to V, but in this case there is a much more natural isomorphism, which does not depend at all on a basis.

6-22 THEOREM. *Let $v \in V$. Then the mapping $e_v : V^* \to k$ defined by $e_v(f)=f(v)$ is a linear transformation from $V^*=\mathrm{Hom}(V,k)$ to k and hence an element of V^{**}. Moreover, the association $v \mapsto e_v$ is an isomorphism from V to V^{**}.*

The map e_v is called *evaluation at v*; this is a fundamental construction in algebra.

PROOF. We leave it as an exercise to show that evaluation at v is linear. Since V and V^{**} have the same dimension, it suffices by the Rank-Nullity Theorem to show that the map $v \mapsto e_v$ has zero kernel. But if e_v is the zero map from V^* to k, then $f(v)=0$ for all $f \in \mathrm{Hom}(V,k)$. We claim that this forces $v=0$. Otherwise, we could extend v to a basis for V, and since *any* set map from a basis to k can be extended to a linear map from V to k, we could certainly construct a linear transformation $f: V \to k$ which does not vanish at v—a clear contradiction. This completes the proof. $\qquad\qquad\qquad\qquad\qquad\qquad\qquad\qquad\qquad\qquad\quad$ $\square$

6.5 Change of Basis

In this final section we analyze what happens to the matrix of a transformation under a change of basis. We shall only discuss linear transformations $T: V \to V$ of a finite-dimensional vector space into itself (these are called *endomorphisms of V*), although the ideas carry over easily to linear maps $T: V \to W$ where W is an arbitrary finite-dimensional space. (See Exercise 25 below for an introduction to this generalization.)

Assume that $\dim(V) = n$ and let both $B = \{v_1,...,v_n\}$ and $B' = \{w_1,...,w_n\}$ be ordered bases for V. Then we have coordinate isomorphisms γ_B and $\gamma_{B'}$ from V to k^n. Let $P \in M_n(k)$ represent the unique isomorphism $(\gamma_B \circ \gamma_{B'}^{-1})$ that makes the following diagram commute:

$$\begin{array}{ccc} & V & \\ \gamma_{B'} \swarrow & & \searrow \gamma_B \\ k^n & \xrightarrow{} & k^n \\ & P & \end{array}$$

In the picture, we identify the matrix P with the linear transformation defined as left multiplication by P (i.e., T_P). Note that since P represents an isomorphism, it is invertible. P is called the *transition matrix from B' to B*.

One can easily give an alternative and less abstract description of P which is often computationally effective. By construction, for $j=1,...,n$

$$P\gamma_{B'}(w_j) = \gamma_B(w_j)$$

The left-hand side is $Pe_j = P^j$, the jth column of P. The right-hand side is the coordinate vector of w_j with respect to the basis B. Thus we have the following result:

RULE. *The jth column of the transition matrix P is the coordinate vector of w_j with respect to the basis B.*

EXAMPLE. Let V equal the space of real polynomials of degree less than or equal to 2. Let $B=\{1,x,x^2\}$ and $B'=\{1+x,1-x,1+x^2\}$. Then the transition matrix from B' to B is evidently

$$P = \begin{pmatrix} 1 & 1 & 1 \\ 1 & -1 & 0 \\ 0 & 0 & 1 \end{pmatrix}$$

We now come to the main point of the section, often called the *change of basis formula*.

6-23 THEOREM. *Let* $T:V\rightarrow V$ *be an endomorphism of a finite-dimensional vector space. Let M denote the matrix of T with respect to the basis B and let N denote the matrix of T with respect to the basis B'. Then*

$$N = P^{-1}MP$$

where P is the transition matrix from B' to B.

PROOF. We give an elegant proof based on the three-dimensional diagram shown in Figure 6.2, which suggests the vertices and edges of a prism.

We see that the triangular faces at either end are commutative since they are precisely the triangles that define the transition matrix P. Moreover, the front (slanted) and back faces are commutative since they are precisely the rectangles that define, respectively, the matrix N of T with respect to the basis B' and the matrix M of T with respect to the basis B. We claim that this forces also the commutativity of the bottom face, which clearly amounts to the matrix equation $PN=MP$. Multiplying both sides of this equality by P^{-1} on the left yields the stated result and completes the proof, subject to verification of our claim.

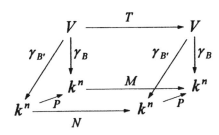

Figure 6.2. Theorem 6-23 may be inferred from the commutativity of this three-dimensional diagram.

PROOF OF CLAIM. This is a lovely diagram chase, and the true flavor is not transmissible in print. (Ask your lecturer for a live demonstration and enjoy the blurred motion of the chalk darting from vertex to vertex as this or that given element is chased across the blackboard. But beware of homological algebraists and algebraic topologists: their performances may have to be video taped and played back in slow motion.) For the present, the following symbolic calculation must suffice. We omit the composition operator:

$$N\gamma_{B'} = \gamma_{B}T \quad \text{(commutativity of the front face)}$$

$$
\begin{aligned}
PN\gamma_{B'} &= P\gamma_{B}T &&\text{(apply } P \text{ on the left to both sides)} \\
&= \gamma_{B}T &&\text{(commutativity of the triangle at the right end)} \\
&= M\gamma_{B} &&\text{(commutativity of the back face)} \\
&= MP\gamma_{B'} &&\text{(commutativity of the triangle at the left end)}
\end{aligned}
$$

$$PN = MP \quad \text{(apply } \gamma_{B'}^{-1} \text{ on the right to both sides)}$$

This establishes the claim. ☐

EXAMPLE. Let V, B, B', and P be as in the previous example. Using the theorem above, we compute the matrix of differentiation on V with respect to B'. We can compute P^{-1} either by directly solving the matrix equation $PX=I_3$ (as described in Section 5.4), or better yet, by finding the transition matrix from B to B'. The latter calculation can be carried out by inspection to yield

$$P^{-1} = \begin{pmatrix} 1/2 & 1/2 & -1/2 \\ 1/2 & -1/2 & -1/2 \\ 0 & 0 & 1 \end{pmatrix}$$

The matrix D of differentiation with respect to B we know from previous examples. Thus the matrix of differentiation with respect to B' is given by

$$P^{-1}DP = \begin{pmatrix} 1/2 & 1/2 & -1/2 \\ 1/2 & -1/2 & -1/2 \\ 0 & 0 & 1 \end{pmatrix} \begin{pmatrix} 0 & 1 & 0 \\ 0 & 0 & 2 \\ 0 & 0 & 0 \end{pmatrix} \begin{pmatrix} 1 & 1 & 1 \\ 1 & -1 & 0 \\ 0 & 0 & 1 \end{pmatrix}$$

Multiplying this out yields

$$P^{-1}DP = \begin{pmatrix} 1/2 & -1/2 & 1 \\ 1/2 & -1/2 & -1 \\ 0 & 0 & 0 \end{pmatrix}$$

The reader should verify the result by direct calculation of the matrix of differentiation relative to B'.

The appearance of the transition matrix in the formula for change of basis suggests the following definition.

DEFINITION. Let $A, B \in M_n(k)$. We say that A is *similar to* B if there exists a matrix $P \in GL_n(k)$ such that $B = P^{-1}AP$. We also write $A \sim B$.

6-24 PROPOSITION. *For all matrices $A, B, C \in M_n(k)$, we have*

(i) $A \sim A$

(ii) $A \sim B \Rightarrow B \sim A$

(iii) $A \sim B$ and $B \sim C \Rightarrow A \sim C$

Hence similarity is an equivalence relation on the set $M_n(k)$.

PROOF. We shall prove only (iii), leaving (i) and (ii) as easy exercises. Suppose that A is similar to B and B is similar to C. Then by definition, there exist matrices $P, Q \in GL_n(k)$ such that

$$B = P^{-1}AP \quad \text{and} \quad C = Q^{-1}BQ$$

Substituting the first equation into the second yields

$$C = Q^{-1}P^{-1}APQ = (PQ)^{-1}A(PQ)$$

which shows that A is similar to C.   □

The connection between similarity and the representation of linear transformations is this:

6-25 PROPOSITION. *Given $A, B \in M_n(k)$, A is similar to B if and only if there exists a vector space V and an endomorphism T of V such that both A and B represent T with respect to (possibly) different bases.*

PROOF. $\Leftarrow$) This is precisely the content of the change of basis formula.

⇒) Suppose that $B=P^{-1}AP$. Let $V=k^n$ so that A represents T_A, left multiplication by the matrix A, with respect to the canonical basis $e_1,...,e_n$. Now consider the basis $P^1,...,P^n$, consisting of the columns of P. (This is indeed a basis by either Theorem 5-6 or Theorem 6-21, since P is invertible.) The transition matrix from $P^1,...,P^n$ to the canonical basis is evidently P itself, so that according to the theorem, the matrix of T_A with respect to this new basis is $B=P^{-1}AP$, as required.   □

Exercises

1. Let $T:\mathbf{R}^2\to\mathbf{R}$ be a linear transformation and suppose that $T(1,1)=5$ and $T(0,1)=2$. Find $T(x_1,x_2)$ for all $x_1,x_2\in\mathbf{R}$. [*Hint*: The vectors $(1,1)$ and $(0,1)$ constitute a basis for $\mathbf{R}^2$.]

2. Find the matrix with respect to the canonical basis of the linear transformation $T:\mathbf{R}^3\to\mathbf{R}^2$ defined by

$$T\begin{pmatrix}x_1\\x_2\\x_3\end{pmatrix}=\begin{pmatrix}5x_1-2x_2+x_3\\4x_2-7x_3\end{pmatrix}$$

3. Let $T:\mathbf{R}^2\to\mathbf{R}^2$ be the linear transformation which is defined by reflection through the line $x=y$. What is $M(T)$, the matrix of T with respect to the canonical basis? Explain geometrically how one knows without calculation that $M(T)^2$ is I_2, the 2×2 identity matrix.

4. Let V be the vector space of real functions spanned by the functions e^x and e^{2x}, which in fact constitute a basis for V. What is the matrix of the differentiation operator on V with respect to this basis?

5. Find the matrix representation (with respect to the canonical basis) for each of the projection maps $p_j:k^n\to k$ defined in Section 3.2.

6. Let V be the real function space spanned by the linearly independent functions $\sin x$, $\sin 2x$, and e^x, which therefore constitute a basis B for V. What is the matrix of the second derivative operator on V with respect to B? (Why does this question make no sense for the first derivative operator?)

7. Let V be a finite-dimensional vector space of dimension n over k. Show that the matrix of the identity map on V with respect to any basis is I_n, the $n\times n$ identity matrix.

8.   Let $T: V \rightarrow V'$ be a linear transformation of vector spaces V and V' with
     bases B and B', respectively. Show that T is invertible if and only if the
     matrix of T with respect to B and B' is likewise invertible. Note that this re-
     sult does not depend on the choice of bases. (*Hint*: Use Proposition 6-13
     and the previous problem.)

9.   Let V be a vector space over k of dimension n. Given $\lambda \in k$, let T_λ denote the
     linear transformation from V to itself defined by $T_\lambda(v) = \lambda v$ for all $v \in V$.
     What is the matrix of T_λ with respect to any basis?

10. Let V be the real vector space of polynomials of degree less than or equal to
     2, with basis $B = \{1, x, x^2\}$. Suppose that for some linear transformation T
     from V to itself, we have that the matrix of T with respect to B is

$$\begin{pmatrix} 2 & 0 & 1 \\ 0 & 1 & 2 \\ 1 & 1 & 1 \end{pmatrix}$$

Compute $T(2 + 5x - 4x^2)$.

11. Let V be the real vector space of polynomials of degree less than or equal to
     2, with basis $B = \{1, x, x^2\}$. Consider the differential operator $D: V \rightarrow V$ de-
     fined by

$$D(p(x)) = 2p(x) + p'(x) - 5p''(x)$$

What is the matrix of D with respect to B? Is D invertible? Why or why
not?

12. Consider $\mathbf{C}$ as a two-dimensional vector space over $\mathbf{R}$ with ordered basis
     $\{1, i\}$. For $z = a + bi$ $(a, b \in \mathbf{R})$, define the mapping $T_z: \mathbf{C} \rightarrow \mathbf{C}$ by

$$T_z(w) = zw$$

so that T_z is just multiplication by z. By the field axioms for the complex
numbers, this is a linear transformation.

(a) What is the matrix of T_z with respect to the given basis?

(b) Use the 2×2 matrix inversion formula (Chapter 5, Exercise 7) to invert the
     matrix associated with a nonzero z. What complex number does this repre-

sent? (Congratulations! You have just rediscovered the well-known formula for the inverse of a complex number.)

(c) What is the matrix associated with the complex number i?

(d) Finally, consider the matrix

$$A = \begin{pmatrix} 0 & -1 \\ 1 & 0 \end{pmatrix}$$

Compute A^{1000} and A^{1003}. (*Hint*: This obviously has something to do with part (c) and should require almost no computation.)

13. Let V be a vector space of dimension 3 and suppose that $B=\{v_1,v_2,v_3\}$ is an ordered basis for V. If the matrix of a linear transformation $T:V \to V$ relative to B is

$$\begin{pmatrix} 2 & 0 & 1 \\ 1 & 1 & 5 \\ 2 & 1 & 2 \end{pmatrix}$$

what is $T(2v_1-v_2+5v_3)$?

14. Let $T:V \to V$ be a linear transformation from a finite-dimensional vector space to itself and suppose that the nullity of T is n. Show that there exists a basis B for V such that the matrix of T with respect to B has precisely n columns of 0's. [*Hint*: If $n=0$, no basis vector is sent to $\mathbf{0}$. If $n>0$, then extend a basis for $\text{Ker}(T)$ to a basis for V.]

15. Let $T:V \to V$ be a linear transformation from a finite-dimensional vector space to itself. Let A be the matrix of T with respect to some basis for V. Show that for any nonnegative integer r, $T^r=0$ if and only if $A^r=0$. (Here the exponents indicate, respectively, iterated composition and iterated multiplication.)

16. Let $T:V \to V$ be a linear transformation from a vector space V of dimension $n \geq 2$ to itself, and suppose that there exists a subspace W of V of dimension m such that $T(W) \subseteq W$ [that is, $T(w) \in W$ for all $w \in W$]. Assume further that $n>m>0$. Show that there is a basis B of V such that the matrix of T relative to B has the form

$$\left(\begin{array}{c|c} P & Q \\ \hline 0 & R \end{array}\right)$$

where P represents an $m \times m$ matrix, Q represents an $m \times (n-m)$ matrix, R represents an $(n-m) \times (n-m)$ matrix, and 0 denotes the $(n-m) \times m$ zero matrix. (*Hint*: Extend a basis for W to a basis for V.)

17. Let V be a vector space of dimension n and suppose that $V = W_1 \oplus W_2$ for subspaces W_1 and W_2 of V of positive dimensions n_1 and n_2, respectively. Now suppose further that T is an endomorphism of V such that $T(W_j) \subseteq W_j$ ($j = 1,2$), so that T maps each of the component subspaces into itself. Show that there is a basis B for V such that the matrix of T with respect to B takes the form

$$\left(\begin{array}{c|c} P & 0 \\ \hline 0 & Q \end{array}\right)$$

where P is an $n_1 \times n_1$ matrix, Q is an $n_2 \times n_2$ matrix, and the 0's represent blocks of zeros of the appropriate sizes.

18. Let V be a finite-dimensional vector space with basis $v_1, \ldots, v_n$ ($n \geq 2$). Let T be the endomorphism of V defined by

$$v_1 \mapsto v_2, v_2 \mapsto v_3, \ldots, v_{n-1} \mapsto v_n, v_n \mapsto v_1$$

Let A be the matrix of T with respect to the given basis. Describe A. What is A^n, the nth power of A?

19. Assume that V is a vector space of dimension n. Let T be an endomorphism of V, and let there be given subspaces $W_1, \ldots, W_n$ of V such that the following three conditions are satisfied:

(i) $W_j \subseteq W_{j+1}$ ($j = 1, \ldots, n-1$)

(ii) $\dim(W_j) = j$ ($j = 1, \ldots, n$)

(iii) $T(W_j) \subseteq W_j$ ($j = 1, \ldots, n$)

Show that there exists a basis B of V such that the matrix of T with respect to B is upper triangular. [Condition (iii) says that each W_j is T-*invariant*, a concept we shall revisit in Chapter 10.]

20. Given the following basis B for $\mathbf{R}^3$, find the transition matrix from B to the canonical basis.

$$B = \{(1,2,-1), (6,0,1), (-1,2,2)\}$$

21. Let V be the vector space of real polynomials of degree less than or equal to 4. Let $B = \{1, x, \dots, x^4\}$ be the usual (ordered) basis of monomials. We can form an alternative basis

$$B' = \{(1+x)^j\}_{j=0,\dots,4}$$

Find the transition matrix from B' to B and the transition matrix from B to B'. (The first is easy; the second slightly more difficult.) What is the relationship between the two transition matrices?

22. Let there be given bases B, B', and B'' for the finite-dimensional vector space V. Suppose that the transition matrix from B'' to B' is Q and that the transition matrix from B' to B is P. State and prove a formula for the transition matrix from B'' to B. (*Hint*: Draw the appropriate commutative diagram.)

23. Show that any two similar matrices have the same rank. (*Hint*: Use Proposition 6-25.)

24. What is the inverse to the map described in Theorem 6-14?

25. Let V and W be finite-dimensional vector spaces with $T \in \mathrm{Hom}(V,W)$. Assume the following additional data:

(i) With respect to the bases B and C for V and W, respectively, the matrix of T is M.

(ii) With respect to the alternative bases B' and C' for V and W, respectively, the matrix of T is N.

(iii) In V, the transition matrix from B' to B is P; in W, the transition matrix from C' to C is Q.

State and prove a formula relating N to M via P and Q. (*Hint*: Revise Figure 6.2 appropriately. The student may agree that this problem is rather more difficult to state than to solve!)

26. Using the ideas of Section 6.4 (in particular, Theorem 6-19), prove without calculation that for matrices A and B, $^t(AB) = {}^tB{}^tA$, whenever these products are defined. [*Hint*: By the contravariance of the dual map,

$$(T_A \circ T_B)^* = T_B^* \circ T_A^*$$

Now write down the matrix of this map in two ways.]

7
Inner Product Spaces

So far we have seen that the definition of an abstract vector space captures the fundamental geometric notion of dimension. There remain, however, at least two other basic geometric ideas that we have not yet addressed: length and angle. To encompass them in the abstract we need to introduce a bit more structure, and in consequence we shall require that our ground field manifest some notion of order. Hence we no longer operate over some abstract field k but rather, for the most part, over the field $\mathbf{R}$ of real numbers. In the final section we shall generalize the results to the complex numbers $\mathbf{C}$.

7.1 Real Inner Product Spaces

DEFINITION. A *real inner product space* V is a real vector space together with a map

$$V \times V \to \mathbf{R}$$
$$(v, w) \mapsto \langle v | w \rangle$$

called a *real inner product*, satisfying the following properties:

(i) $\langle v | v \rangle \geq 0 \ \forall v \in V$, with equality if, and only if, $v = 0$

(ii) $\langle v | w \rangle = \langle w | v \rangle \ \forall v, w \in V$

(iii) $\langle u + v | w \rangle = \langle u | w \rangle + \langle v | w \rangle \ \forall u, v, w \in V$

   $\langle av | w \rangle = a \langle v | w \rangle \ \forall v, w \in V, \ a \in \mathbf{R}$

Note that according to (ii), the relations expressed in (iii) hold equally well on the other side.

These properties are called, respectively, *positive definiteness*, *symmetry*, and *bilinearity*. The term bilinearity reflects the linearity of the inner product in either variable; that is, for all $v_0 \in V$, the following maps are linear:

$$V \to \mathbf{R} \qquad\qquad V \to \mathbf{R}$$
$$v \mapsto \langle v \,|\, v_0 \rangle \qquad\quad v \mapsto \langle v_0 \,|\, v \rangle$$

EXAMPLES

(1) Let $V = \mathbf{R}^n$. Then we define the *canonical inner product* or *dot product* on V by the formula

$$\langle x | y \rangle = \sum_{j=1}^{n} x_j y_j$$

The requisite properties are evident from the ordinary arithmetic properties of real numbers. One should also notice that this inner product can be realized as matrix multiplication:

$$\langle x | y \rangle = {}^t x y$$

where we use the transpose operator to convert x from a column vector to a row vector.

(2) Let $V = \mathscr{C}^0([a,b])$, the vector space of real-valued continuous functions on the closed interval $[a,b]$. We define an inner product on V by the formula

$$\langle f | g \rangle = \int_a^b f(x) g(x) \, dx$$

The requisite properties follow at once from the familiar rules for the definite integral.

Notice that this second example is not so distant from the first. A function is in some sense a vector with infinitely many coordinates, one for each point x of the domain. We cannot directly add the infinitely many products of corresponding components $f(x)g(x)$ as we do in the case of $\mathbf{R}^n$, but we can replace the suggested calculation by a limit process operating on discrete summations: the definite integral.

An inner product immediately yields a reasonable definition of length.

DEFINITION. Let V be an inner product space. Then given a vector $v \in V$, we define the *length* or *norm* of v, denoted $|v|$, by

$$|v| = \langle v | v \rangle^{1/2}$$

A vector of length 1 is called a *unit vector*.

Length has two obvious properties that follow at once from the definition of an inner product:

(i) $|v|=0 \Leftrightarrow v=0, \quad \forall v \in V$

(ii) $|av|=|a|\cdot|v| \quad \forall v \in V, a \in \mathbf{R}$

Note that in (ii), the vertical bars around a real number denote the ordinary absolute value function; around a vector they denote length or norm as here defined.

EXAMPLES REVISITED

(1) For our first example above, we have

$$|x|=\left(\sum_{j=1}^{n} x_j^{\,2}\right)^{1/2}$$

which agrees in small dimensions with our ordinary notion of length, via the Pythagorean Theorem.

(2) For our second example above, we have

$$|f|=\left(\int_a^b f(x)^2 dx\right)^{1/2}$$

This notion of length for functions is not *a priori* geometric (and most emphatically has nothing to do with arc length), but the formalism does yield reasonable geometric interpretations and a good deal of useful and sometimes surprising information.

We now come to the fundamental inequality for inner products. The proof given here for real inner product spaces is efficient but does not generalize to the complex case. A better, more conceptual proof awaits us in Section 7.3.

7-1 THEOREM. (The Cauchy-Schwarz Inequality) *Let V be an inner product space. Then for all vectors v and w in V,*

$$|\langle v|w\rangle| \leq |v|\cdot|w|$$

Note carefully that the vertical bars on the left signify absolute value; on the right they signify the length function. (In light of this ambiguity, some denote length $\|v\|$, but in practice our simpler notation causes no confusion.)

REMARK. As we shall soon see, almost all of the results of this and the following section hold for both real and complex inner product spaces. Whenever this is so, we state our assertions simply for inner product spaces, implicitly including both cases.

PROOF. For all $x \in \mathbf{R}$,

$$\langle v+xw | v+xw \rangle \geq 0$$

Using bilinearity to expand the left side, we find that

$$\langle v|v \rangle + 2x \langle v|w \rangle + x^2 \langle w|w \rangle \geq 0$$

or, equivalently,

$$|w|^2 x^2 + 2 \langle v|w \rangle x + |v|^2 \geq 0$$

Now considering this as a statement about a polynomial in the indeterminate x, we conclude that this polynomial has at most one real root, so that the discriminant is less than or equal to 0. Therefore

$$4 \langle v|w \rangle^2 - 4|v|^2 |w|^2 \leq 0$$

and

$$\langle v|w \rangle^2 \leq |v|^2 |w|^2$$

Taking square roots, we have the stated inequality.   □

7-2 COROLLARY. (The Triangle Inequality) *For all v and w in V,*

$$|v+w| \leq |v| + |w|$$

PROOF. This is an exercise in bilinearity with an opportune use of the Cauchy-Schwarz Inequality:

$$\begin{aligned}
|v+w|^2 &= \langle v+w|v+w \rangle \\
&= \langle v|v \rangle + 2\langle v|w \rangle + \langle w|w \rangle \\
&\leq |v|^2 + 2|v||w| + |w|^2 \\
&\leq (|v|+|w|)^2
\end{aligned}$$

Again taking square roots of both sides, we obtain the desired result.   □

The student should plot two nonparallel vectors in $\mathbf{R}^2$ together with their sum to see how this result gets its name: since the shortest distance between two points in the Euclidean plane is a straight line, the sum of the lengths of two legs of a triangle must exceed the length of the third.

We see now that the Cauchy-Schwarz Inequality strengthens our sense that the abstract length function associated with an inner product does indeed behave like ordinary length. It is also the key to extending another primitive geometric idea.

DEFINITION. Let v and w be nonzero vectors in the inner product space V. Then the *angle between v and w* is the number $\theta \in [0, \pi]$ defined by the equation

$$\cos \theta = \frac{\langle v | w \rangle}{|v| \|w\|}$$

Note that this definition makes sense since the quotient on the right is guaranteed to lie between -1 and $+1$ by the Cauchy-Schwarz Inequality. (Physics students may recognize the definition of the dot product as given in many texts: $x \cdot y = |x| |y| \cos \theta$. Of course this cannot be a definition of the dot product for us, since we are using essentially the same equation to define θ.)

As an important special case of the angle between two vectors we have the following definition:

DEFINITION. Let v and w be vectors in the inner product space V. Then we say v *is orthogonal to w* and write $v \perp w$ if $\langle v | w \rangle = 0$.

For nonzero vectors in a real inner product space, this, of course, corresponds to the case where the angle between v and w is $\pi/2$.

A family $v_1, \ldots, v_m$ of nonzero vectors in V is called an *orthogonal family* if it satisfies the condition that $\langle v_i | v_j \rangle = 0$ whenever $i \neq j$. An orthogonal family consisting entirely of unit vectors is called an *orthonormal family*. In this case

$$\langle v_i | v_j \rangle = \delta_{ij} \quad (1 \leq i, j \leq m)$$

(We may, of course, also speak of an orthogonal or orthonormal set.)

EXAMPLES

(1) The canonical basis for $\mathbf{R}^n$ is an orthonormal family and hence called an *orthonormal basis* for $\mathbf{R}^n$. There are infinitely many others. For example, in the special case $n = 2$, we can obtain the alternative orthonormal basis

$$\left(\frac{\sqrt{2}}{2}, \frac{\sqrt{2}}{2} \right), \left(-\frac{\sqrt{2}}{2}, \frac{\sqrt{2}}{2} \right)$$

by rotating the canonical basis vectors counterclockwise about the origin by $\pi/4$ radians. Other angles, of course, yield other alternatives.

(2) The student can show (with some patience or perhaps a table of integrals) that the following infinite family of functions in $\mathscr{C}^0([-\pi,+\pi])$ is orthogonal:

$$1, \cos x, \sin x, \cos 2x, \sin 2x, \cos 3x, \sin 3x,\ldots$$

They are not, however, orthonormal, but by a simple scalar adjustment one can make them so. (See Exercise 15 below.)

In light of our previous work, these examples might suggest that an orthogonal family is linearly independent. This is indeed the case, as we shall now show.

7-3 PROPOSITION. *Every orthogonal family is also a linearly independent family.*

PROOF. Let $v_1,\ldots,v_m$ constitute an orthogonal family and suppose that

$$\sum_{j=1}^m a_j v_j = 0$$

Then taking the inner product of the summation with any member v_k of the family and using bilinearity and orthogonality, we find that

$$0 = \langle \sum_{j=1}^m a_j v_j | v_k \rangle = \sum_{j=1}^m a_j \langle v_j | v_k \rangle = a_k \langle v_k | v_k \rangle$$

Since $\langle v_k | v_k \rangle$ is nonzero (an inner product is positive definite), it follows at once that $a_k = 0$. Since this is true for all k, the assertion follows. $\square$

We next look at another consequence of orthogonality. This is a generalization of what many consider to be the most important theorem in mathematics.

7-4 THEOREM. (The Pythagorean Theorem) *Suppose that the vectors* $v_1,\ldots,v_m$ *constitute an orthogonal family in the inner product space V. Then*

$$|\sum_{j=1}^m v_j|^2 = \sum_{j=1}^m |v_j|^2$$

REMARK. Consider what this theorem says for two orthogonal vectors a and b in the Euclidean plane $\mathbf{R}^2$. Let the scalars a and b denote the corresponding lengths. Their sum c represents the hypotenuse of a right triangle, the length of which we denote c. The formula above then asserts that $c^2 = a^2 + b^2$, and this is indeed just the statement of the Pythagorean Theorem that we learned in high school geometry.

PROOF. This is an almost trivial calculation:

$$|\sum_{j=1}^{m} v_j|^2 = \langle \sum_{j=1}^{m} v_j | \sum_{j=1}^{m} v_j \rangle$$

$$= \sum_{1 \leq j,k \leq m} \langle v_j | v_k \rangle$$

$$= \sum_{j=1}^{m} \langle v_j | v_j \rangle$$

$$= \sum_{j=1}^{m} |v_j|^2$$

The point is that only the diagonal terms contribute on the second line, since the vectors in question are assumed pairwise orthogonal.   ❑

Before leaving this basic discussion, we should mention one last elementary fact concerning the connection between inner products and a certain class of matrices. A matrix $C \in M_n(\mathbf{R})$ is called *positive definite* if $^t x C x > 0$ for all non-zero x in $\mathbf{R}^n$. According to Theorem 6-21, since left multiplication by a positive definite matrix clearly has kernel 0 ($Cx=0$ without $x=0$ would certainly violate the defining condition), a positive definite matrix must be invertible.

7-5 PROPOSITION. *Let V be a finite-dimensional real vector space of dimension n with (ordered) basis B. Then every real inner product on V arises in the form*

$$\langle v | w \rangle = {}^t \gamma_B(v) C \gamma_B(w)$$

where C is a real $n \times n$ positive definite, symmetric matrix.

PROOF. Given a positive definite, symmetric matrix C, it is easy to show that the formula above defines an inner product. Conversely, given any real inner product on V, one can associate the symmetric matrix $C=(c_{ij})$ defined by

$$c_{ij} = \langle v_i | v_j \rangle \quad (1 \leq i,j \leq n)$$

where $v_1,...,v_n$ is the given basis. One can show by direct calculation that the given inner product is equivalent to the formula and that therefore C is indeed positive definite.   ❑

7.2 Orthogonal Bases and Orthogonal Projection

Throughout, let V be a real inner product space. (All of the results proven here will also hold for complex inner product spaces.)

DEFINITION. An orthonormal family in V that is also a basis is called an *orthonormal basis* for V.

We have seen, for example, that the canonical basis for $\mathbf{R}^n$ is an orthonormal basis. Orthonormal bases are particularly easy to work with since finding coordinates and lengths is trivial.

7-6 PROPOSITION. *Let $u_1,\ldots,u_n$ be an orthonormal basis for V. Then for all v in V, the following assertions hold:*

(i) $v = \sum_{j=1}^{n} \langle v|u_j \rangle u_j$

(ii) $|v|^2 = \sum_{j=1}^{n} \langle v|u_j \rangle^2$

PROOF. Let $v = a_1 u_1 + \cdots + a_n u_n$. Then

$$\langle v|u_k \rangle = \langle \sum_{j=1}^{n} a_j u_j | u_k \rangle = a_k \langle u_k|u_k \rangle = a_k$$

as claimed. (Since the u_j are in particular orthogonal, only the term for which the subscripts match contributes to the summation.) The second statement is just the Pythagorean Theorem: the u_j are moreover unit vectors, so the length of each term $\langle v|u_j \rangle u_j$ appearing in part (i) is exactly the absolute value of the scalar coefficient $\langle v|u_j \rangle$.   □

The summands in part (i) are especially important and give rise to yet another fundamental definition with strong geometric connotations.

DEFINITION. Let v lie in V and let $u \in V$ be a unit vector. Then

$$\mathrm{pr}_u(v) = \langle v|u \rangle u$$

is called the *orthogonal projection of v onto u.* More generally, if W is a subspace of V with orthonormal basis $u_1,...,u_m$, then

$$\mathrm{pr}_W(v) = \sum_{j=1}^{m} \langle v|u_j\rangle u_j$$

is called the *orthogonal projection of v onto W*. Note that by construction the orthogonal projection lies in W since it is a linear combination of elements of W. (For convenience, we define the projection of any vector onto the zero subspace to be **0**.)

Figure 7.1 should help the student to visualize a low-dimensional case of orthogonal projection onto a subspace. In the picture, the vector v is shown above the planar subspace W, while its orthogonal projection $\mathrm{pr}_W(v)$ is, of course, shown to lie in W. Other incidental features of the illustration are formalized below in Lemma 7-7 and Corollary 7-11.

With this new language we may paraphrase the previous proposition by saying that a vector is the sum of its orthogonal projections onto the elements of a finite orthonormal basis. Moreover, we shall show shortly (Theorem 7-8) that every finite-dimensional inner product space does in fact admit an orthonormal basis, and therefore the orthogonal projection of a vector onto a finite-dimensional subspace is always defined. We shall deduce further that this projection does not depend upon the particular choice of an orthonormal basis (Corollary 7-10) and thus depends only upon W itself. This justifies our terminology in speaking of the projection onto W without explicit reference to a basis.

7-7 LEMMA. *In the context of the definition above, for all v in V*

$$(v-\mathrm{pr}_W(v)) \perp u_k \quad (k=1,...,m)$$

and hence $v-\mathrm{pr}_W(v)$ is orthogonal to every vector in W.

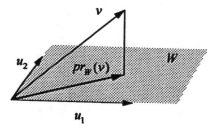

Figure 7.1. The orthogonal projection of v onto the subspace W spanned by the orthonormal set u_1 and u_2.

PROOF. Again a straightforward calculation:

$$\langle v - \mathrm{pr}_W(v)|u_k\rangle = \langle v|u_k\rangle - \langle \mathrm{pr}_W(v)|u_k\rangle$$

$$= \langle v|u_k\rangle - \langle \sum_{j=1}^{m}\langle v|u_j\rangle u_j|u_k\rangle$$

$$= \langle v|u_k\rangle - \sum_{j=1}^{m}\langle v|u_j\rangle\langle u_j|u_k\rangle$$

$$= \langle v|u_k\rangle - \langle v|u_k\rangle\langle u_k|u_k\rangle = 0$$

The transition from line 3 to line 4 is justified by the orthogonality of the u's; the last step requires that the u's be unit vectors.

The second assertion of the lemma follows from bilinearity, since every element of W can be written as a linear combination of the orthonormal basis vectors.   ☐

We are now ready to prove the existence of orthonormal bases for finite-dimensional inner product spaces. The proof actually exhibits an algorithm for constructing such a basis, albeit a tedious one to execute by hand.

7-8 THEOREM. (The Gram-Schmidt Orthonormalization Process) *Every finite-dimensional inner product space V has an orthonormal basis.*

PROOF. The proof is constructive. If V is the zero space, the empty set is vacuously an orthonormal basis. Otherwise, let $v_1,\ldots,v_n$ be a basis for V and define an ascending chain of subspaces W_j in V by $W_j = \mathrm{Span}(v_1,\ldots,v_j), j=1,\ldots,n$. Thus

$$W_1 \subseteq W_2 \subseteq \ldots \subseteq W_n = V$$

Now set

$$u_1 = \frac{v_1}{|v_1|}$$

Then u_1 is a unit vector and thus an orthonormal basis for W_1. We can therefore define orthogonal projections onto W_1, and thus define a second unit vector

$$u_2 = \frac{v_2 - \mathrm{pr}_{W_1}(v_2)}{|v_2 - \mathrm{pr}_{W_1}(v_2)|}$$

The point is that we subtract from v_2 its orthogonal projection onto W_1 to obtain a vector orthogonal to u_1, and then we normalize it. This yields an orthonormal basis for W_2. (Clearly two orthogonal, hence linearly independent, vectors in the two-dimensional space W_2 must be a basis.) Iterating this procedure, at stage k we define

$$u_k = \frac{v_k - \mathrm{pr}_{W_{k-1}}(v_k)}{|v_k - \mathrm{pr}_{W_{k-1}}(v_k)|}$$

for $k=2,\ldots,n$ and thus obtain an orthonormal basis $u_1,\ldots,u_n$ for $W_n=V$.   □

EXAMPLES

(1) We apply Gram-Schmidt to the basis for $\mathbf{R}^3$ consisting of the vectors

$$(2,0,0), (1,5,0), (1,2,2)$$

One finds at once that $u_1=(1,0,0)$. To find u_2, we first subtract from $(1,5,0)$ its orthogonal projection onto the subspace spanned by u_1:

$$(0,5,0) = (1,5,0) - 1\cdot(1,0,0)$$

We normalize this to obtain $u_2 = (0,1,0)$. Finally, we subtract from $(1,2,2)$ its orthogonal projection onto the subspace spanned by u_1 and u_2:

$$(0,0,2) = (1,2,2) - 1\cdot(1,0,0) - 2\cdot(0,1,0)$$

We again normalize and consequently find that $u_3=(0,0,1)$. We have thus recovered the canonical basis! [Do not be misled—things don't always work out this easily, as you will see by repeating the procedure on these same vectors in reverse order, starting with $(1,2,2)$. Try it.]

(2) A more interesting problem is to apply Gram-Schmidt to the three-dimensional subspace V of $\mathscr{C}^0([-1,+1])$ spanned by the functions 1, x, and x^2. After tedious but routine calculations, one obtains the functions

$$\frac{\sqrt{2}}{2}, \quad \frac{\sqrt{6}}{2}x, \quad \text{and} \quad \frac{\sqrt{10}}{4}(3x^2 - 1)$$

This is the start of something big: the so-called *Legendre polynomials*, which are so important in mathematical physics.

Orthogonal Complementation

Let V be a finite-dimensional inner product space with subspace W. We define $W^{\perp}$, the *orthogonal complement of W*, by

$$W^{\perp} = \{v \in V : \langle v|w \rangle = 0 \ \forall w \in W\}$$

Thus $W^{\perp}$ consists of all vectors in V orthogonal to every vector in W.

7-9 PROPOSITION. *In the context of the preceding definition, the following assertions hold:*

(i) $W^{\perp}$ *is a subspace of V.*

(ii) $W \cap W^{\perp} = \{0\}$.

(iii) *For all $v \in V$, there exist unique $w \in W$ and $w^{\perp} \in W^{\perp}$ such that $v = w + w^{\perp}$; moreover, w is precisely the orthogonal projection of v onto W.*

Thus in the language of direct sums, $V = W \oplus W^{\perp}$.

PROOF. Part (i) is an exercise. For (ii), note that if w lies in both sets, it must be orthogonal to itself. Thus $\langle w|w \rangle = 0$ and w must be 0. For (iii), given a vector v, define w to be $\mathrm{pr}_W(v)$ (computed against any orthonormal basis for W) and $w^{\perp}$ to be $v - w$. Then clearly $v = w + w^{\perp}$. Uniqueness follows from Proposition 3-8.   ❏

7-10 COROLLARY. *The orthogonal projection $\mathrm{pr}_W(v)$ is independent of the choice of orthonormal basis for W.*

PROOF. This is essentially the uniqueness statement of the preceding result.   ❏

7-11 COROLLARY. *For all v, $\mathrm{pr}_W(v)$ is the point of W closest to v.*

Here closest means with respect to the obvious distance function $|v - w|$, the length of the difference of two vectors.

PROOF. Let $v = w + w^{\perp}$ as in the previous proposition, so that w is the orthogonal projection of v onto W. Then for any $w' \in W$, by the Pythagorean Theorem

$$|w' - v|^2 = |(w' - w) - w^{\perp}|^2$$
$$= |w' - w|^2 + |w^{\perp}|^2$$

and this is clearly minimized when $w' = w$.   ❏

7.3 Complex Inner Product Spaces

In this brief section we generalize the basic definitions associated with a real inner product to the complex numbers **C**. Most of the previous discussion applies with little or no change.

First recall that every complex number is uniquely expressible in the form

$$z = a + bi$$

where a and b are real numbers and $i^2 = -1$. We call a the *real part* of z, denoted Re(z), and b the *imaginary part* of z, denoted Im(z). The *complex conjugate* of z, denoted $\bar{z}$, is defined as follows:

$$\bar{z} = a - bi$$

That is, we change the sign of the imaginary part. Thus conjugation has no effect on real numbers. It is easy to show that complex conjugation is both additive and multiplicative, so that

$$\overline{(z+w)} = \bar{z} + \bar{w}$$

$$\overline{(zw)} = \bar{z}\,\bar{w}$$

Accordingly, one says that conjugation is an *automorphism* of the field **C**. Finally, note that for all complex numbers z,

$$|z| = \sqrt{z\bar{z}} = \sqrt{a^2 + b^2}$$

With these preliminaries, we can now state the definition of a complex inner product space.

DEFINITION. A *complex inner product space* V is a complex vector space together with a map

$$V \times V \to \mathbf{C}$$
$$(v, w) \mapsto \langle v | w \rangle$$

called a *complex inner product*, satisfying the following properties:

(i) $\langle v | v \rangle \geq 0 \ \forall v \in V$, with equality if and only if $v = 0$.

(ii) $\langle v | w \rangle = \overline{\langle w | v \rangle} \ \forall v, w \in V$

(iii) $\langle u+v|w\rangle=\langle u|w\rangle + \langle v|w\rangle \ \forall u,v,w\in V$

$\quad \langle av|w\rangle=a\langle v|w\rangle \ \forall v,w\in V, \ a\in\mathbf{C}$

It is implicit in (i) that $\langle v|v\rangle$ is real for all $v\in V$. From (ii) and (iii), we obtain an additional property (called *antilinearity*) in the second variable:

(iv) $\langle u|v+w\rangle=\langle u|v\rangle + \langle u|w\rangle \ \forall u,v,w\in V$

$\quad \langle v|aw\rangle= \bar{a}\langle v|w\rangle \ \forall v,w\in V, \ a\in\mathbf{C}$

EXAMPLES

(1) Let $V=\mathbf{C}^n$. Then we define the *canonical inner product* or *dot product* on V by the formula

$$\langle x|y\rangle=\sum_{j=1}^{n}x_j\bar{y}_j$$

The requisite properties follow from the ordinary arithmetic properties of complex numbers. (See Exercise 26 below for some important remarks on this definition.)

(2) Let V be the space of complex-valued continuous functions on the closed interval $[a,b]$. We define an inner product on V by the formula

$$\langle f|g\rangle=\int_{a}^{b}f(x)\overline{g(x)}\,dx$$

The requisite properties follow from the rules for the definite integral. (Note: The integration of a continuous complex-valued function f defined on an interval $[a,b]$ is achieved by separately integrating, in the usual way, both its real and imaginary parts. Hence the familiar rules apply.)

The definition of length for a complex inner product space is identical to that for a real inner product space; that is,

$$|v|= \sqrt{\langle v|v\rangle}$$

The key to showing that length as defined here has the expected properties (and later the key to defining the angle between two vectors) is again the Cauchy-Schwarz Inequality.

7-12 THEOREM. (The Cauchy-Schwarz Inequality) *Let V be an inner product space (real or complex). Then for all v and w in V,*

$$|\langle v|w\rangle| \le |v|\cdot|w|$$

PROOF. The simple proof given previously, which made essential use of the symmetry of the real inner product, will not go through in this case. We must give a more intrinsic argument. Assume that v is not the zero vector, since otherwise the assertion is trivial. Define

$$\lambda = \frac{\langle v|w\rangle}{|w|^2}$$

and

$$v' = v - \lambda w$$

In light of Section 7.2, λw is formally the orthogonal projection of v onto the subspace spanned by w, and it follows as previously that $\langle v'|w\rangle = 0$. Now expand the square of the length of v':

$$\begin{aligned}
0 \le |v'|^2 &= \langle v'|v - \lambda w\rangle \\
&= \langle v'|v\rangle \\
&= \langle v - \lambda w|v\rangle \\
&= |v|^2 - \langle \lambda w|v\rangle
\end{aligned}$$

Thus

$$\lambda\langle w|v\rangle \le |v|^2$$

and substituting the definition of λ into this last inequality yields

$$\frac{\langle v|w\rangle\langle w|v\rangle}{|w|^2} \le |v|^2$$

$$\langle v|w\rangle\langle w|v\rangle \le |v|^2|w|^2$$

Now since $\langle w|v\rangle$ is the complex conjugate of $\langle v|w\rangle$, the product on the left is precisely $|\langle v|w\rangle|^2$. Taking square roots of both sides, the inequality is established. ∎

As a corollary, we have the Triangle Inequality also for complex inner product spaces; the previous proof goes through with one small change.

7-13 COROLLARY. (The Triangle Inequality) *Let V be an inner product space (real or complex). For all v and w in V,*

$$|v+w| \le |v| + |w|$$

PROOF. Again we use the linearity of the inner product and the Cauchy-Schwarz Inequality, but in this case we must beware that the cross terms are not equal, but rather conjugate:

$$\begin{aligned}
|v+w|^2 &= \langle v+w \,|\, v+w \rangle \\
&= \langle v|v \rangle + \langle v|w \rangle + \langle w|v \rangle + \langle w|w \rangle \\
&= \langle v|v \rangle + 2\,\mathrm{Re}(\langle v|w \rangle) + \langle w|w \rangle \\
&\le |v|^2 + 2|v\|w| + |w|^2 \\
&= (|v|+|w|)^2
\end{aligned}$$

In passing from the third to the fourth line, we make use of the following chain of inequalities:

$$|\mathrm{Re}(\langle v|w \rangle)| \le |\langle v|w \rangle| \le |v| \cdot |w|$$

Taking square roots of both sides yields the stated result.   ❏

We can now define the angle between nonzero vectors v and w in a complex inner product space to be the unique number $\theta \in [0, \pi]$ defined by the equation

$$\cos \theta = \frac{\mathrm{Re}(\langle v|w \rangle)}{|v\|w|}$$

This makes sense because the quotient lies between -1 and $+1$, according to the Cauchy-Schwarz Inequality. (To verify this, we appeal to the same chain of inequalities exhibited in the previous proof.)

Orthogonality and orthonormality are defined as above, and the proof that an orthogonal set is linearly independent applies without change, as does the proof of the Pythagorean Theorem. (For some important observations on the relationship between angle and orthogonality in a complex inner product space, see Exercises 29 and 30 below.)

Finally, note that all of the results of Section 7.2 are valid, with no changes in the proofs, for complex inner product spaces. In particular, we may apply Gram-Schmidt to show that every finite-dimensional complex inner product space has an orthonormal basis.

Exercises

1. In $\mathbf{R}^3$, compute the inner product of $(1,2,-1)$ and $(2,1,4)$. What is the length of each vector? What is the angle between these vectors?

2. What is the angle between the vectors $(1,2,4)$ and $(2,5,1)$ in $\mathbf{R}^3$? You may leave your answer in terms of the inverse cosine function.

3. Find all vectors in $\mathbf{R}^3$ which are orthogonal to both of the following vectors:

$$(1,2,0) \text{ and } (1,0,1)$$

This amounts to a homogeneous system of two equations in three unknowns. (Cognoscenti of the vector cross product may have their own ideas.)

4. Compute the inner product $\langle f | g \rangle$ in $\mathscr{C}^0([-\pi,+\pi])$ for the following functions:

$$f(x) = 2x \text{ and } g(x) = \sin x$$

5. In the context of the previous problem, find the length of the functions f and g. What is the angle between these functions? Interpret the Cauchy-Schwarz Inequality in this special case.

6. In the inner product space $\mathscr{C}^0([-1,+1])$, for which n are the monomials x^n orthogonal to the constant function 1?

7. In the context of the previous problem, what is the length of each of the monomials x^n?

8. Use the Cauchy-Schwarz Inequality on the appropriate inner product space to bound the definite integral

$$\int_0^{\pi/2} \sqrt{x \sin x}\, dx$$

9. Prove that in an n-dimensional inner product space an orthogonal family may contain at most n vectors. (*Hint:* If you need more than two sentences for this one, you've missed the point.)

10. Carry out the Gram-Schmidt orthonormalization process on the following pair of vectors in $\mathbf{R}^2$ to obtain an orthonormal basis:

$$(2,1) \text{ and } (-1,3)$$

11. Apply the Gram-Schmidt orthonormalization process to the vectors $(3,4,5)$ and $(1,0,1)$ to obtain an orthonormal pair of vectors with the same span.

12. Let $V=\mathbf{R}^3$ and let W be the subspace of V spanned by the vectors $(1,0,1)$ and $(0,1,0)$. What point of W is closest to the vector $(6,2,5)$?

13. In $\mathbf{R}^3$, let W be the subspace spanned by the vectors $(1,1,2)$ and $(1,1,-1)$. What point of W is closest to the vector $(4,5,-2)$?

14. In $\mathbf{R}^3$, find the orthogonal projection of $(2,2,5)$ on the subspace spanned by the vectors $(2,1,1)$ and $(0,2,1)$. (*Hint*: First apply Gram-Schmidt to the spanning set.)

15. Granting that the functions

$$1, \cos x, \sin x, \cos 2x, \sin 2x,\dots$$

constitute an orthogonal family in $\mathscr{C}^0([-\pi,+\pi])$, modify each function by a scalar (not necessarily the same one) to convert this to an orthonormal family.

16. Let $f \in V=\mathscr{C}^0([-\pi,+\pi])$. Give a formula for the orthogonal projection of f onto the subspace of V spanned by the $2n+1$ functions 1, $\cos x$, $\sin x$, $\cos 2x$, $\sin 2x,\dots,$ $\cos nx$, $\sin nx$. Expand this in terms of the appropriate definite integrals, which you need not (and cannot!) compute. (Thus begins the development of *Fourier series* or *harmonic analysis*.)

17. Let v and w be vectors in the inner product space V, with w nonzero. Write down a formula for the orthogonal projection of v onto the subspace spanned by w. (Remember that w is not necessarily a unit vector.) This is often called simply the *projection of v onto w*.

18. Let W be the subspace spanned by $(1,1,1)$ in $\mathbf{R}^3$. Find a basis for $W^{\perp}$, the orthogonal complement of W. [*Hint*: Any two linearly independent vectors orthogonal to $(1,1,1)$ will do.]

19. Let W be a finite-dimensional subspace of the inner product space V. Show that the projection map $\mathrm{pr}_W : V \to V$ is a linear transformation. What are the kernel and image of this map? (*Hint*: Use Proposition 7-9.)

20. In the context of the previous problem, what is $\text{pr}_{W^\circ}\text{pr}_W$?

21. Let W be a nontrivial, proper subspace of the finite-dimensional inner product space V and again consider the projection map $\text{pr}_W: V \to V$. Assume that the dimension of W is m. Show that there is a basis B for V such that the matrix of pr_W with respect to B takes the form

$$\left(\begin{array}{c|c} I_m & 0 \\ \hline 0 & 0 \end{array} \right)$$

where I_m is the $m{\times}m$ identity matrix, and the zeros represent zero matrices of appropriate sizes. (*Hint:* Consider appending a basis for $W^\perp$ to a basis for W.)

22. Show that complex conjugation is a bijective map from $\mathbf{C}$ into itself. (*Hint:* What is the inverse map?)

23. Show that $\mathbf{C}^n$ is isomorphic to $\mathbf{R}^{2n}$ as real vector spaces. [*Hint:* Use the functions $\text{Re}(z)$ and $\text{Im}(z)$ defined on the complex numbers.]

24. Find the length of the following vector in $\mathbf{C}^2$:

$$(2+5i, \ 1-4i)$$

Remember that the canonical inner product on $\mathbf{C}^2$ requires conjugation.

25. Let V be the complex vector space of continuous complex-valued functions on the interval $[-\pi, +\pi]$. Consider the function $f \in V$ defined by $f(x) = e^{ix}$. Find the length of f in V. (*Hint:* The complex conjugate of e^{ix} is e^{-ix} and the formal properties of exponents still apply to the complex exponential.)

26. In $\mathbf{C}^n$ we defined the canonical inner product by

$$\langle x|y \rangle = \sum_{j=1}^{n} x_j \bar{y}_j$$

Why not define it instead more simply as

$$\langle x|y \rangle = \sum_{j=1}^{n} x_j y_j$$

without conjugation of the y_j? What essential feature would be lost?

27. Show that in a complex inner product space we have

$$|av| = |a| \cdot |v| \quad \forall v \in V, \, a \in \mathbf{C}$$

This generalizes another familiar property of real inner products and is needed to extend the proof of Proposition 7-6 to the complex case.

28. Let V be a complex vector space with basis $\{v_j\}_{j \in J}$, where J is some index set, possibly infinite. Then show that V *as a real vector space* has the basis $\{v_j, iv_j\}_{j \in J}$. Hence if V has dimension n over $\mathbf{C}$, it has dimension $2n$ over $\mathbf{R}$.

29. Let V be a complex inner product space and let v be any nonzero vector in V. Show that $\langle v | iv \rangle \neq 0$ but nevertheless the angle between v and iv is $\pi/2$. Hence in a complex inner product space, vectors forming a right angle need not be formally orthogonal. How, then, can one reconcile these notions? See the following problem.

30. Show that for two complex numbers z and w, the product $z\overline{w}$ is purely imaginary if and only if z and w are orthogonal as points of $\mathbf{R}^2$. [For this, identify the complex number $a+bi$ with the point (a,b) in $\mathbf{R}^2$.] This at least reconciles perpendicularity and orthogonality in $\mathbf{C}$. Now generalize this to higher dimensions.

8
Determinants

The determinant is an amazing function that in some sense measures the invertibility of an $n \times n$ matrix. In this chapter we first give a formal, functional description of the determinant, showing its existence via a recursive definition. We then exhibit a direct formula, which leads at once to a statement of uniqueness and a surprising multiplicative property. [The determinant turns out to be a group homomorphism from $GL_n(k)$ to k^*.] Finally, from this multiplicativity one easily deduces that a square matrix is invertible if and only if its determinant is not zero.

8.1 Existence and Basic Properties

We begin with a handy bit of notation. Let k be a field. If $A \in M_n(k)$, $n > 1$, then henceforth

$$\partial_{ij} A \in M_{n-1}(k) \quad (1 \le i, j \le n)$$

is the $(n-1) \times (n-1)$ matrix which results upon the deletion of the ith row and jth column of A. For example,

$$\partial_{12} \begin{pmatrix} a_{11} & a_{12} & a_{13} \\ a_{21} & a_{22} & a_{23} \\ a_{31} & a_{32} & a_{33} \end{pmatrix} = \begin{pmatrix} a_{21} & a_{23} \\ a_{31} & a_{33} \end{pmatrix}$$

This operation turns out to be critical to the sequel.

As further preparation for the main theorem of this chapter, next recall that we often write a matrix A as the amalgamation of its columns; that is,

$$A = (A^1, \ldots, A^n)$$

where A^j is the jth column of A.

8-1 THEOREM. (The Fundamental Theorem of Determinants) *For each $n \geq 1$, there exists a unique map*

$$\det : M_n(k) \to k$$

called the determinant, *satisfying the following rules:*

(i) MULTILINEARITY. *Let $A = (A^1, \ldots, A^n)$ and suppose that $A^j = \lambda_1 C_1 + \lambda_2 C_2$ for column vectors $C_1, C_2 \in k^n$ and scalars λ_1, λ_2. Then*

$$\det(A) = \lambda_1 \det(A^1, \ldots, \underset{\substack{\uparrow \\ \text{column} j}}{C_1}, \ldots, A^n) + \lambda_2 \det(A^1, \ldots, \underset{\substack{\uparrow \\ \text{column} j}}{C_2}, \ldots, A^n)$$

That is, the determinant is linear in every column.

(ii) ALTERNATION OF SIGN. *Suppose that $A = (A^1, \ldots, A^n)$ and that $A^j = A^{j+1}$ for some j, so that two adjacent columns of A are identical. Then*

$$\det(A) = 0$$

(iii) NORMALIZATION. *The determinant of the $n \times n$ identity matrix is the unity of k; that is,*

$$\det(I_n) = 1$$

for all n.

We shall see shortly how property (ii) earns its name.

PROOF OF EXISTENCE. (Uniqueness will be shown later as a deeper consequence of the properties established here.) We begin by giving a recursive definition of the determinant.

A 1×1 matrix is just an element of the ground field k, so for $n = 1$ define $\det(a) = a$. All three properties clearly hold—the second vacuously. For $n > 1$, define $\det : M_n(k) \to k$ recursively by the formulas

$$\det(A) = \sum_{j=1}^{n} (-1)^{j+1} a_{1j} \det(\partial_{1j} A) \tag{8.1}$$

(This is called *expansion by the first row*. Examples are coming!) Thus the evaluation of the determinant for a given matrix proceeds in terms of matrices of smaller sizes until we finally reach the scalar case, for which the map is given explicitly. Properties (i), (ii), and (iii) are now verified by induction.

(i) For notational simplicity we give the argument in the first column, but it clearly applies to any column. So assume that $A^1 = \lambda_1 C_1 + \lambda_2 C_2$, and, in particular, that $a_{11} = \lambda_1 c_1 + \lambda_2 c_2$ where c_1 and c_2 are, respectively, the first entries of C_1 and C_2. Then by the definition above (Eq. 8.1),

$$\det(A) = (\lambda_1 c_1 + \lambda_2 c_2) \det(\partial_{11} A) + \sum_{j=2}^{n} (-1)^{j+1} a_{1j} \det(\partial_{1j} A)$$

and so by induction,

$$\det(A) = \lambda_1 c_1 \det(\partial_{11} A) + \lambda_2 c_2 \det(\partial_{11} A)$$
$$+ \sum_{j=2}^{n} (-1)^{j+1} a_{1j} \lambda_1 \det(\partial_{1j}(C_1, A^2, ..., A^n))$$
$$+ \sum_{j=2}^{n} (-1)^{j+1} a_{1j} \lambda_2 \det(\partial_{1j}(C_2, A^2, ..., A^n))$$

But notice now that

$$\partial_{11} A = \partial_{11}(C_1, A^2, ..., A^n) = \partial_{11}(C_2, A^2, ..., A^n)$$

since none of the three matrices involves the disputed first column of A. Thus we may combine the first and third and second and fourth summands of our prior expansion to obtain

$$\det(A) = \lambda_1 \left\{ c_1 \det(\partial_{11}(C_1, A^2, ..., A^n)) + \sum_{j=2}^{n} (-1)^{j+1} a_{1j} \det(\partial_{1j}(C_1, A^2, ..., A^n)) \right\}$$
$$+ \lambda_2 \left\{ c_2 \det(\partial_{11}(C_2, A^2, ..., A^n)) + \sum_{j=2}^{n} (-1)^{j+1} a_{1j} \det(\partial_{1j}(C_2, A^2, ..., A^n)) \right\}$$

Observe next that $c_1 \det(\partial_{11}(C_1, A^2, ..., A^n))$ is just the first term in the expansion of $\det(C_1, A^2, ..., A^n)$ and that $c_2 \det(\partial_{11}(C_2, A^2, ..., A^n))$ is similarly the first term in the expansion of $\det(C_2, A^2, ..., A^n)$. Hence we can absorb these terms into the adjacent summations to obtain

$$\det(A) = \lambda_1 \det(C_1, A^2, ..., A^n) + \lambda_2 \det(C_2, A^2, ..., A^n)$$

as claimed.

(ii) We now demonstrate the second property in the case of the first two columns. As above, the argument carries over to the general case without difficulty. Suppose that $A^1 = A^2$. Then by definition

$$\det(A) = a_{11} \det(\partial_1 A) - a_{12} \det(\partial_2 A) + \sum_{j=3}^{n} (-1)^{j+1} a_{1j} \det(\partial_j A)$$

and the first two terms cancel since $a_{11} = a_{12}$ and $\partial_{11} A = \partial_{12} A$. The remaining terms involve the matrices $\partial_{1j} A$ ($j = 3, \ldots, n$) for which again columns 1 and 2 are identical. Hence, by induction, all of their determinants are zero, and the result follows. (The case $n = 2$ is covered directly by the expansion above.)

(iii) Normalization is immediate: Since the $(1,1)$-entry of I_n is the only nonzero entry on the first row, our recursive formula for the determinant yields

$$\det(I_n) = 1 \cdot \det(\partial_{11} I_n) = \det(I_{n-1})$$

which is 1 by induction. This completes the proof of the theorem.   ❑

EXAMPLES

We give two examples of our recursive definition, both of which yield useful nonrecursive formulas.

(1) Calculation of the determinant by expansion of the first row in the 2×2 case yields at once the formula

$$\det \begin{pmatrix} a & b \\ c & d \end{pmatrix} = ad - bc$$

Recall that this expression arose in Chapter 5, Exercise 7, in connection with the invertibility of 2×2-matrices.

(2) Using the 2×2-formula and expansion by the first row, we can easily deduce a formula for the determinant of the general 3×3 matrix $A = (a_{ij})$:

$$\det(A) = a_{11} \det \begin{pmatrix} a_{22} & a_{23} \\ a_{32} & a_{33} \end{pmatrix} - a_{12} \det \begin{pmatrix} a_{21} & a_{23} \\ a_{31} & a_{33} \end{pmatrix} + a_{13} \det \begin{pmatrix} a_{21} & a_{22} \\ a_{31} & a_{32} \end{pmatrix}$$

Applying the previous result to each of the terms, this resolves itself to

$$\det(A) = a_{11}(a_{22}a_{33} - a_{32}a_{23}) - a_{12}(a_{21}a_{33} - a_{31}a_{23}) + a_{13}(a_{21}a_{32} - a_{31}a_{22})$$

$$= a_{11}a_{22}a_{33} + a_{12}a_{23}a_{31} + a_{13}a_{21}a_{32} - a_{31}a_{22}a_{13} - a_{32}a_{23}a_{11} - a_{33}a_{21}a_{12}$$

This formula can be remembered since it is generated by a simple pattern of diagonals. (Find it!)

If we repeat this analysis for the general 4×4 matrix, what do we find? The determinant resolves itself into four 3×3 determinants, each of which involves six terms. Hence we might deduce a 24-term formula, of no particular value. (The 3×3 pattern does not persist!) Arguing inductively, one sees that recursive expansion of the determinant for the general $n{\times}n$ matrix involves the computation of $n!$ terms. Since this expression grows explosively, the method is clearly not computationally effective. Nonetheless, the appearance of the factorial in this context is suggestive and foreshadows things to come.

8-2 COROLLARY. *The determinant has the following additional properties:*

(i) *Suppose that A' is obtained from A by interchange of two adjacent columns. Then*

$$\det(A') = -\det(A)$$

(ii) *Suppose that A' is obtained from A by interchange of any two columns. Then*

$$\det(A') = -\det(A)$$

(iii) *Suppose that any two columns of A are identical. Then*

$$\det(A) = 0$$

PROOF. (i) Let $A = (A^1, \ldots, A^j, A^{j+1}, \ldots, A^n)$ and let $A' = (A^1, \ldots, A^{j+1}, A^j, \ldots, A^n)$. Now consider the matrix

$$A'' = (A^1, \ldots, A^j + A^{j+1}, A^j + A^{j+1}, \ldots, A^n)$$

Since A'' has two identical adjacent columns, its determinant is 0. Thus by multilinearity we have

$$0 = \det(A^1, \ldots, A^j, A^j, \ldots, A^n) + \det(A^1, \ldots, A^j, A^{j+1}, \ldots, A^n)$$
$$+ \det(A^1, \ldots, A^{j+1}, A^j, \ldots, A^n) + \det(A^1, \ldots, A^{j+1}, A^{j+1}, \ldots, A^n)$$

The first and fourth terms in the sum are zero for having identical adjacent columns, while the second and third are, respectively, $\det(A)$ and $\det(A')$. Thus

$$0 = \det(A) + \det(A')$$

and the result follows.

(ii) Recall from the theory of permutations (see the proof of Lemma 1-7) that any transposition is the result of an odd number of adjacent transpositions. Therefore the exchange of any two columns of A can be accomplished by an odd number of adjacent swaps, each of which changes the sign of the determinant once. An odd number of sign changes amounts to a single sign change, as claimed.

(iii) If any two columns of A are identical, consider what happens when we swap these columns. On the one hand, the matrix A changes not at all, whence its determinant likewise remains unchanged. On the other hand, according to part (ii) of the present corollary, the determinant changes sign. But the only possibility for a number unaffected by a sign change is zero! ◻

This corollary explains why we described property (ii) of Theorem 8-1 as alternation of sign. The next result begins to forge a link between the determinant and the theory of vector spaces and linear transformations. (The full connection is given in Theorem 8-10 below.)

8-3 COROLLARY. *Suppose that the columns of A are linearly dependent. Then* $\det(A)=0$.

PROOF. If the columns of A are linearly dependent, then one column is a linear combination of the others. By multilinearity, we can expand the determinant into a sum of determinants of matrices with at least two identical columns. To be more precise, if

$$A^j = \sum_{k \neq j} \lambda_k A^k$$

then

$$\det(A) = \sum_{k \neq j} \lambda_k \det(\underset{\substack{\uparrow \\ \text{column } j}}{A^1, \ldots, A^k}, \ldots, A^n)$$

But part (iii) of the previous result implies that each summand, and hence the total determinant, is zero. ◻

Further progress in the theory of determinants depends upon a complementary description of this map (one that could have served as an alternative definition, albeit an ugly one). This is given in the following section.

8.2 A Nonrecursive Formula; Uniqueness

We next give a nonrecursive formula for the determinant; this is notationally compact, but practically intractable except in the smallest dimensions. Its real importance is theoretical, as we shall see shortly. The reader should first review the definition of S_n, the symmetric group on n letters, and $\sigma: S_n \to \{\pm 1\}$, the sign homomorphism, in Section 1.5.

8-4 THEOREM. *For all $A \in M_n(k)$,*

$$\det(A) = \sum_{\pi \in S_n} \sigma(\pi) a_{\pi(1)1} \cdots a_{\pi(n)n}$$

where π varies over the symmetric group on n letters and σ denotes the sign homomorphism.

Note that there are $n!$ terms in the summation. This is consistent with our earlier analysis.

PROOF. Let E_j denote the $n \times 1$ matrix with 1 in the jth component, 0 elsewhere. Then we may write A as

$$(a_{11}E_1 + \cdots + a_{n1}E_n, A^2, \ldots, A^n)$$

By multilinearity,

$$\det(A) = \sum_{i=1}^{n} a_{i1} \det(E_i, A^2, \ldots, A^n)$$

Repeating this argument in the second column, we find further that

$$\det(A) = \sum_{i=1}^{n} \sum_{j=1}^{n} a_{i1} a_{j2} \det(E_i, E_j, \ldots, A^n)$$

Repeating this over all n columns, we obtain all possible combinations of coefficients and column vectors E_j. This may be expressed as

$$\det(A) = \sum_{\varphi} a_{\varphi(1)1} \cdots a_{\varphi(n)n} \det(E_{\varphi(1)}, \ldots, E_{\varphi(n)})$$

where φ ranges over all functions $\{1, \ldots, n\} \to \{1, \ldots, n\}$. But by part (iii) of the Corollary 8-2,

$$\det(E_{\varphi(1)},\ldots,E_{\varphi(n)}) = 0$$

if $\varphi(i)=\varphi(j)$ for any distinct pair of indices i,j. Thus rather than summing over *all* of the functions $\varphi:\{1,\ldots,n\}\rightarrow\{1,\ldots,n\}$, it suffices to sum over only those that are injective. But by the Pigeonhole Principle, this is just the set of permutations of $\{1,\ldots,n\}$. Hence

$$\det(A) = \sum_{\pi \in S_n} a_{\pi(1)1}\cdots a_{\pi(n)n} \det(E_{\pi(1)},\ldots,E_{\pi(n)})$$

Finally consider the determinants that appear in this expression. We know that we can obtain $(E_{\pi(1)},\ldots,E_{\pi(n)})$ from $I_n=(E_1,\ldots,E_n)$ by successive transpositions, the required number of such being even or odd according to the sign of the permutation π. Therefore

$$\det(E_{\pi(1)},\ldots,E_{\pi(n)}) = \sigma(\pi)\det(E_1,\ldots,E_n) = \sigma(\pi)\det(I_n) = \sigma(\pi)$$

and this establishes the formula.   □

8-5 COROLLARY. *For all $A \in M_n(k)$, $\det(A)=\det({}^tA)$; that is, the determinant of a matrix is equal to the determinant of its transpose.*

PROOF. The proof reduces to a calculation that uses two elementary facts about permutations:

(i) as π varies over S_n, so does π^{-1} (the inversion map on a group is its own inverse and hence is a bijection);

(ii) π and π^{-1} have the same sign (Chapter 2, Exercise 17).

With this in mind, we have

$$\det(A) = \sum_{\pi \in S_n} \sigma(\pi)a_{\pi(1)1}\cdots a_{\pi(n)n}$$

$$= \sum_{\pi \in S_n} \sigma(\pi^{-1})a_{\pi^{-1}(1)1}\cdots a_{\pi^{-1}(n)n}$$

$$= \sum_{\pi \in S_n} \sigma(\pi)a_{1\pi(1)}\cdots a_{n\pi(n)}$$

$$= \det({}^tA)$$

and this completes the proof.   □

Uniqueness of the Determinant

The uniqueness of the determinant map is an immediate consequence of Theorem 8-4, as we shall now demonstrate.

PROOF OF UNIQUENESS. Our nonrecursive formula depends solely on the properties of multilinearity, alternation of sign, and normality. Thus any mapping $M_n(k) \to k$ which satisfies these three properties must also satisfy this closed formula. Hence there can be only one such map. $\square$

We shall see next, as a consequence of uniqueness, that the recursive definition of the determinant given in the Fundamental Theorem in terms of the first row applies equally well to any other row or column. This is a great convenience for matrices with rows or columns consisting of many zeroes.

8-6 PROPOSITION. (Expansion by Rows and Columns) *The following formulas for the determinant also hold:*

(i) *For any fixed row index i, $1 \le i \le n$,*

$$\det(A) = \sum_{j=1}^{n} (-1)^{i+j} a_{ij} \det(\partial_{ij} A)$$

(ii) *For any fixed column index j, $1 \le j \le n$,*

$$\det(A) = \sum_{i=1}^{n} (-1)^{i+j} a_{ij} \det(\partial_{ij} A)$$

PROOF. (i) The proof of the Fundamental Theorem applies to any row. Hence by uniqueness, expansion by any row must yield the same result.

(ii) Since a matrix and its transpose have the same determinant, expansion by any given column is the same as expansion by the corresponding row of the transpose matrix. Hence the result follows from part (i). $\square$

8-7 COROLLARY. *The determinant of a triangular matrix is the product of its diagonal entries.*

PROOF. For a lower triangular matrix, this follows at once by iterative expansion in the first row. For upper triangular matrices, it follows at once by iterative expansion in the first column. (In the latter case, one could argue alternatively that an upper triangular matrix is the transpose of a lower triangular matrix; both matrices have the same diagonal entries and the same determinant.) $\square$

We have also a second corollary, which will be much needed in Chapter 10 in connection with the reduction of matrices to certain standard forms.

8-8 COROLLARY. *Let $A \in M_n(k)$ have the form*

$$A = \left(\begin{array}{c|c} P & Q \\ \hline 0 & R \end{array}\right)$$

where P represents an $m \times m$ matrix, Q represents an $m \times (n-m)$ matrix, R represents an $(n-m) \times (n-m)$ matrix, and 0 denotes the $(n-m) \times m$ zero matrix. Then

$$\det(A) = \det(P)\det(R)$$

Note in particular that the entries in Q play no part whatsoever in the evaluation of the determinant.

PROOF. We argue by induction on m, the size of P. If $m = 1$, then expansion by the first column yields only one term

$$\det(A) = a_{11}\det(\partial_{11}A)$$

which reduces immediately to the formula given, since in this case $P = a_{11}$ while $R = \partial_{11}A$. Suppose next that $m > 1$. Expanding again by the first column, we make the following three-step calculation, explained below:

$$\det(A) = \sum_{i=1}^{m}(-1)^{i+1}a_{i1}\det(\partial_{i1}A)$$

$$= \sum_{i=1}^{m}(-1)^{i+1}a_{i1}\det(\partial_{i1}P)\det(R)$$

$$= \det(P)\det(R)$$

In the first step, we limit our expansion to only the first m terms because the remaining entries in A's first column are by assumption zero. In the second step, we invoke the induction hypothesis: the matrices $\partial_{i1}A$ ($i \leq m$) are all of the same form as A, but the top-left block is one size smaller, having lost both a row and a column. (Note that R is unruffled by these deletions.) In the third step, we merely factor out $\det(R)$ and then recognize the remaining summation as the first-column expansion of $\det(P)$. This completes the proof. ☐

8.3 The Determinant of a Product; Invertibility

The nonrecursive formula for the determinant may be used to reveal yet another remarkable aspect of this map which seems completely foreign to its previous formal properties.

8-9 THEOREM. *Let A and B lie in $M_n(k)$. Then*

$$\det(AB) = \det(A) \cdot \det(B)$$

PROOF. This is similar to the proof of Theorem 8-4. The key is to recall that

$$AB = (A \cdot B^1, \ldots, A \cdot B^j, \ldots, A \cdot B^n)$$

and that the jth column may be expressed as

$$A \cdot B^j = b_{1j} A^1 + \cdots + b_{nj} A^n$$

Using this fact, we first expand the determinant of AB by multilinearity to obtain

$$\det(AB) = \sum_{\varphi} b_{\varphi(1)1} \cdots b_{\varphi(n)n} \det(A^{\varphi(1)}, \ldots, A^{\varphi(n)})$$

where the sum is taken over all functions $\varphi : \{1,\ldots,n\} \to \{1,\ldots,n\}$. As previously, the only functions to contribute are the permutations, and this yields

$$\det(AB) = \sum_{\pi \in S_n} b_{\pi(1)1} \cdots b_{\pi(n)n} \det(A^{\pi(1)}, \ldots, A^{\pi(n)})$$

Each summand then involves the determinant of A following some permutation π of its columns. But this reduces to the determinant of A times the sign of π. Factoring out $\det(A)$ and appealing to the nonrecursive formula, we have

$$\det(AB) = \det(A) \sum_{\pi \in S_n} \sigma(\pi) b_{\pi(1)1} \cdots b_{\pi(n)n} = \det(A)\det(B)$$

as claimed.   □

   The multiplicative nature of the determinant once again extends our characterization of invertibility. (The student will be relieved to know that this theorem has now reached full maturity, at least as far as this text is concerned.)

8-10 THEOREM. *Let $A \in M_n(k)$. Then the following twelve statements are equivalent:*

(i)   *The linear system $Ax=y$ has at least one solution for all $y \in k^n$.*

(ii)   *The columns of A span k^n.*

(iii) *The rows of A span k^n.*

(iv)   *The homogeneous linear system $Ax=0$ has only the trivial solution $x=0$.*

(v)   *The columns of A are linearly independent.*

(vi)   *The rows of A are linearly independent.*

(vii) *The linear system $Ax=y$ has exactly one solution for all $y \in k^n$.*

(viii) *The columns of A constitute a basis for k^n.*

(ix)   *The rows of A constitute a basis for k^n.*

(x)   *A is invertible; i.e., $A \in GL_n(k)$.*

(xi)   *The determinant of A is nonzero.*

(xii) *The determinant of ${}^t A$ is nonzero.*

PROOF. We have already established the equivalence of (i) through (x) in Theorem 6-21. Moreover, we saw above that if A has linearly dependent columns, then the determinant of A is zero, showing that (xi) implies (v). The equivalence of (xi) and (xii) is immediate from Corollary 8-5, leaving us finally to show that (x) implies (xi). This is easy:

Suppose that A is invertible. Then there exists a matrix B such that $AB=I_n$. But then

$$\det(A) \det(B) = \det(AB) = \det(I_n) = 1$$

and clearly $\det(A)$ is nonzero, completing the proof.   □

The following corollary is immediate from the equivalence of (x) and (xi) and the multiplicativity of the determinant.

8-11 COROLLARY. *The map $\det : GL_n(k) \to k^*$ is a homomorphism from the group of invertible $n \times n$ matrices to the group of nonzero elements of k.* □

The kernel of $\det : GL_n(k) \to k^*$ is by definition the set of all $n \times n$ matrices of determinant 1. This much celebrated object is called the *special linear group* and denoted $SL_n(k)$.

REMARK. Our proof that a matrix is invertible if and only if it has nonzero determinant is not optimal in the sense that we have implicitly used that a vector space over a field is free. In fact, one does not need so much structure, and an analogous result might have been proven for a mere commutative ring k with unity. [$A \in GL_n(k)$ if and only if $\det(A)$ is invertible in k.] The proof is based on *Cramer's Rule*, a theoretically attractive, but computationally impractical formula for solving linear systems via determinants.

Exercises

1. Using the recursive definition given in the proof of Theorem 8-1 (i.e., expansion by the first row), systematically evaluate the determinant of the following matrix:

$$A = \begin{pmatrix} 1 & 2 & 1 \\ 0 & 1 & 1 \\ 1 & 0 & 2 \end{pmatrix}$$

2. For any angle θ, evaluate the determinant of the matrix

$$M(\theta) = \begin{pmatrix} \cos\theta & -\sin\theta \\ \sin\theta & \cos\theta \end{pmatrix}$$

Recall that this is the matrix of rotation about the origin by the angle θ.

3. Show that for $A \in M_n(k)$ and $\lambda \in k$, $\det(\lambda A) = \lambda^n \det(A)$.

4. Show that if a matrix has a row or column of 0's, then its determinant is 0. (*Hint*: Make an elegant appeal to multilinearity.)

5. Show that the determinant of a matrix is unchanged if we add a scalar multiple of one column to another. (Since according to Corollary 8-5 the determinant of a matrix is equal to the determinant of its transpose, the same is also true for rows.)

6. Consider the matrix

$$A = \begin{pmatrix} x_1 & a_1 & b_1 \\ x_2 & a_2 & b_2 \\ x_3 & a_3 & b_3 \end{pmatrix}$$

in $M_3(k)$, where the a's and b's are fixed and the x's may vary. Show that the set of all (x_1,x_2,x_3) such that $\det(A)=0$ is a subspace of k^3 of dimension at least 2. (*Hint*: Recall that the determinant is a linear transformation in each column.)

7. Evaluate the determinant of the matrix

$$A = \begin{pmatrix} 0 & 1 & 0 & 0 & 0 \\ 0 & 0 & 1 & 0 & 0 \\ 0 & 0 & 0 & 0 & 1 \\ 1 & 0 & 0 & 0 & 0 \\ 0 & 0 & 0 & 1 & 0 \end{pmatrix}$$

This is an example of a *permutation matrix* since it acts by permuting the canonical basis vectors of k^5.

8. Evaluate the determinant of the following matrix using expansion by any row or column as appropriate.

$$A = \begin{pmatrix} 1 & 1 & -1 & 2 \\ 0 & 1 & 2 & 0 \\ 4 & 0 & 3 & 1 \\ 0 & 2 & 0 & 0 \end{pmatrix}$$

9. Evaluate the determinant of the following matrix using expansion by any row or column as appropriate.

$$A = \begin{pmatrix} 0 & 1 & -1 & 2 \\ 1 & 1 & 2 & 0 \\ 4 & 0 & 2 & -1 \\ 3 & 2 & 0 & 4 \end{pmatrix}$$

10. Evaluate the determinant of the following matrix *without* expanding by either rows or columns.

$$\begin{pmatrix} 1 & 1 & 5 \\ 2 & 0 & 2 \\ 4 & 0 & 0 \end{pmatrix}$$

(*Hint*: Try swapping columns to reach a triangular matrix; be sure to keep track of the sign.)

11. Evaluate the determinant of the matrix

$$A = \begin{pmatrix} 2 & -1 & 3 & 1 \\ 0 & 2 & 2 & 0 \\ 3 & 0 & 1 & 0 \\ 0 & 0 & 2 & 0 \end{pmatrix}$$

12. Evaluate the determinant of the matrix

$$A = \begin{pmatrix} 2 & 6 & 1 & 8 & 0 \\ 0 & 4 & 5 & 7 & 1 \\ 0 & 0 & 4 & 9 & 7 \\ 0 & 0 & 0 & 1 & 5 \\ 0 & 0 & 0 & 1 & 0 \end{pmatrix}$$

13. Consider a matrix $A = (a_{ij}) \in M_n(k)$ whose elements are all zero above the *minor diagonal* (from bottom left to top right). Give a succinct formula for the determinant of A. (*Hint*: Work out the 2×2, 3×3, 4×4, and 5×5 cases explicitly and then generalize.)

14. Let $A \in M_n(k)$ have factorization $A = LU$ into the product of a lower triangular matrix L and an upper triangular matrix U. Show that the determinant of A is equal to the product of the diagonal terms in both L and U.

15. Let $A, B \in M_n(k)$ and suppose that A is singular (i.e., not invertible). Show that the product AB is also singular. [*Hint*: Use the determinant. What is $\det(AB)$?]

16. Suppose that A is an invertible matrix such that both A and A^{-1} consist entirely of integers. Show that the determinant of A is either $+1$ or -1. (*Hint*: the determinant of a matrix of integers must also be an integer; this follows at once from the nonrecursive formula.)

17. Consider each of the elementary row operations defined on matrices in Section 5.3. Analyze how each affects the determinant of a square matrix.

18. For all $A \in M_n(\mathbf{R})$, prove that $\det(A\,{}^tA) \geq 0$.

19. Prove that if A is similar to B, then $\det(A) = \det(B)$. (See Section 6.5 for the definition of similarity.)

20. Prove that the following matrices are *not* similar:

$$A = \begin{pmatrix} 1 & -1 & 0 \\ 0 & 2 & 5 \\ 0 & 0 & 3 \end{pmatrix} \qquad B = \begin{pmatrix} 2 & 0 & 0 \\ -1 & 4 & 0 \\ 0 & 3 & 7 \end{pmatrix}$$

(*Hint*: The previous problem might help.)

21. Determine whether the following set of vectors is linearly independent. (You now have an easy way to do this without resorting to a linear system.)

$$(1,2,-1), \ (6,0,2), \ (4,-4,2)$$

22. Show that

$$\det \begin{pmatrix} 1 & x_1 & x_1^2 \\ 1 & x_2 & x_2^2 \\ 1 & x_3 & x_3^2 \end{pmatrix} = (x_2 - x_1)(x_3 - x_1)(x_3 - x_2)$$

Conclude that the vectors

$$(1,x_1,x_1^2), \ (1,x_2,x_2^2), \ (1,x_3,x_3^2)$$

are linearly independent if and only if x_1, x_2, and x_3 are distinct. Generalize this to higher dimensions. (*Hint*: Use Exercise 5 above.)

23. Let $f_1,\dots,f_n$ be a family of functions in $\mathscr{C}^\infty(\mathbf{R})$. Note that if

$$\sum_{j=1}^{n} \lambda_j f_j(x) = 0$$

for some family of coefficients $\lambda_j \in \mathbf{R}$, then

$$\sum_{j=1}^{n} \lambda_j f_j^{(k)}(x) = 0$$

for all $k \geq 0$. (As usual, $f^{(k)}$ denotes the kth derivative of f.) With this background, show that if $f_1,\ldots,f_n$ are linearly dependent, then

$$
\det \begin{pmatrix}
f_1 & f_2 & \cdots & f_n \\
f_1^{(1)} & f_2^{(1)} & \cdots & f_n^{(1)} \\
\vdots & \vdots & & \vdots \\
f_1^{(n-1)} & f_2^{(n-1)} & \cdots & f_n^{(n-1)}
\end{pmatrix} = 0
$$

(This is called the *Wronskian determinant* of the family $f_1,\ldots,f_n$ and is critical in the theory of differential equations.)

24. Use the previous exercise to show that the functions e^x, e^{2x}, and e^{3x} are linearly independent; that is, show that

$$
\det \begin{pmatrix}
e^x & e^{2x} & e^{3x} \\
e^x & 2e^{2x} & 3e^{3x} \\
e^x & 4e^{2x} & 9e^{3x}
\end{pmatrix} \neq 0
$$

25. Use Exercises 22 and 23 to show that the collection of functions

$$
e^{\lambda_1 x}, e^{\lambda_2 x}, \ldots, e^{\lambda_n x} \quad (\lambda_j \in \mathbf{R})
$$

is linearly independent if and only if the numbers λ_j are distinct. This generalizes the previous exercise.

26. Let

$$
A = \begin{pmatrix}
1 & 0 & 1 \\
0 & 2 & 1 \\
1 & 4 & 3
\end{pmatrix}
$$

Find A^{-1} or prove that it does not exist.

27. Show that $\mathrm{SL}_n(k)$ is a subgroup of $\mathrm{GL}_n(k)$. (*Hint:* You need only reread the definition of the special linear group.)

28. Consider the 2×2 matrix

$$A = \begin{pmatrix} a & b \\ c & d \end{pmatrix}$$

Let $P=(a,c)$ and $Q=(b,d)$ be the points in $\mathbf{R}^2$ defined by the columns of A. These points span a parallelogram whose vertices are $\mathbf{0}$, P, Q, and $P+Q$. Find the area of this object. (*Hint*: Find the area of the rectangle and each of the four triangles in the Figure 8.1.)

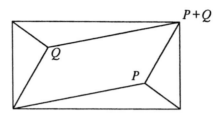

Figure 8.1. The vectors P and Q in $\mathbf{R}^2$ span a parallelogram.

9
Eigenvalues and Eigenvectors

This chapter introduces and, to a limited extent, solves one of the classical problems associated with linear processes: their decomposition into well-behaved, independent component subprocesses. What is especially noteworthy and exciting about the material is that it uses all of the major concepts introduced so far, including the representation of linear transformations, real and complex inner product spaces, and the theory of determinants. The final theorem of Section 9.3 is as exquisite as any work of art.

9.1 Definitions and Elementary Properties

DEFINITION. Let $T: V \rightarrow V$ be a linear transformation on a vector space V over k, and suppose that for some $\lambda \in k$ there exists a nonzero $v \in V$ such that

$$T(v) = \lambda v$$

Then λ is called an *eigenvalue of T*, and every vector v (including 0) that satisfies the equation above is called an *eigenvector belonging to λ*.

The terms *characteristic value* and *characteristic vector* are common synonyms for eigenvalue and eigenvector. Note that geometrically an eigenvector of T is a nonzero vector v such that $T(v)$ is parallel to v.

EXAMPLES

(1) Let $V = \mathbf{R}^2$ and let

$$A = \begin{pmatrix} 0 & 1 \\ 1 & 0 \end{pmatrix}$$

We have the linear transformation $T_A: V \rightarrow V$ defined by left multiplication by A. Since

$$T\begin{pmatrix} +1 \\ +1 \end{pmatrix} = +1 \cdot \begin{pmatrix} +1 \\ +1 \end{pmatrix}, \quad T\begin{pmatrix} +1 \\ -1 \end{pmatrix} = -1 \cdot \begin{pmatrix} +1 \\ -1 \end{pmatrix}$$

both +1 and −1 are eigenvalues of T_A (or simply of A) with corresponding eigenvectors as given.

(2) Consider differentiation as a linear transformation on the real vector space $\mathscr{C}^\infty(\mathbf{R})$, the set of infinitely differentiable functions from $\mathbf{R}$ to $\mathbf{R}$. Every number $\lambda \in \mathbf{R}$ is an eigenvalue for differentiation with corresponding eigenvector $Ce^{\lambda x}$, where C is an arbitrary real constant.

As an easy exercise, the student should show that, in general, the set of eigenvectors belonging to a given eigenvalue λ constitutes a subspace of V. This is called the *eigenspace* belonging to λ and generalizes the notion of kernel (which may be considered the eigenspace belonging to 0).

A basis for V consisting of eigenvectors of T is called an *eigenbasis for V with respect to T*. The matrix of a transformation with respect to an eigenbasis takes an especially simple form.

9-1 THEOREM. *Let $T: V \to V$ be a linear transformation on a finite-dimensional vector space. Then T is representable by a diagonal matrix if and only if there exists an eigenbasis for V with respect to T. In this case, the diagonal entries are precisely the eigenvalues of T.*

Such a transformation is called *diagonalizable*.

PROOF. Suppose that $v_1, \dots, v_n$ is a basis B of eigenvectors and assume that the corresponding eigenvalues are $\lambda_1, \dots, \lambda_n$. Then

$$T(v_j) = \lambda_j v_j$$

and the matrix of T with respect to B is indeed the diagonal matrix whose diagonal entries are $\lambda_1, \dots, \lambda_n$. Conversely, if the matrix of T with respect to a given basis is diagonal, then the corresponding basis vectors satisfy equations of the form above and are therefore an eigenbasis for V with respect to T. $\square$

Recall from Section 6.5 that if B and B' are bases for the finite-dimensional vector space V, then for any given endomorphism T of V,

$$M_{B'}(T) = P^{-1}M_B(T)P$$

where P is the transition matrix from B' to B. Recall, too, that matrices related by an equation of this form are called *similar*. The previous theorem then has the following interpretation:

9-2 COROLLARY. *$A \in M_n(k)$ is similar to a diagonal matrix if and only if there exists an eigenbasis for k^n with respect to A.* □

EXAMPLES

(1) Continuing with Example 1 above, we see that V has an eigenbasis $(1,1)$, $(1,-1)$ with respect to T. Note that these vectors are also orthogonal. This is no mere coincidence, as we shall see shortly.

(2) Consider the linear transformation $T_\theta : \mathbf{R}^2 \to \mathbf{R}^2$ defined by the matrix

$$\begin{pmatrix} \cos\theta & -\sin\theta \\ \sin\theta & \cos\theta \end{pmatrix}$$

As we have seen before, this is counterclockwise rotation by θ around the origin. Clearly T_θ has no eigenvalues if θ is not an integral multiple of π, since in this case no nonzero vector can be rotated onto a scalar multiple of itself. Thus there is no eigenbasis and the matrix is not diagonalizable.

The question naturally arises, how does one find eigenvalues and eigenvectors, if any? Surprisingly, this can be framed as a problem in finding roots of polynomials. We first treat the case of a matrix acting as a linear transformation on k^n. We shall see shortly that this suffices for the general case of an arbitrary endomorphism of a finite-dimensional vector space.

DEFINITION. Let $A \in M_n(k)$ and let t be an indeterminate. Then the nth degree polynomial

$$p(t) = \det(tI_n - A)$$

is called the *characteristic polynomial of A*.

9-3 THEOREM. *The eigenvalues of A are the roots in k of the characteristic polynomial of A.*

PROOF. Let λ lie in the ground field k. Then $p(\lambda) = \det(\lambda I_n - A)$, and we have the following chain of equivalences:

$$\det(\lambda I_n - A) = 0 \Leftrightarrow \lambda I_n - A \text{ is singular (recall that this means non-invertible)}$$
$$\Leftrightarrow \exists x \in k^n, x \neq 0, \text{ such that } (\lambda I_n - A)x = 0$$
$$\Leftrightarrow \exists x \in k^n, x \neq 0, \text{ such that } Ax = \lambda x \qquad \square$$

The proof shows something more: given an eigenvalue λ, the problem of finding the corresponding eigenspace reduces to solving the homogeneous linear system $(\lambda I_n - A)x = 0$.

Note also how this result depends upon the ground field. The polynomial $t^2 + 1$ has no roots in $\mathbf{R}$, but does have roots in $\mathbf{C}$.

EXAMPLES

(1) Again consider

$$A = \begin{pmatrix} 0 & 1 \\ 1 & 0 \end{pmatrix}$$

We compute the characteristic polynomial of A:

$$p(t) = \det(tI_2 - A) = \det\begin{pmatrix} t & -1 \\ -1 & t \end{pmatrix} = t^2 - 1$$

We see that the roots are $+1$ and -1, precisely the eigenvalues of A that we had observed previously.

(2) Next consider

$$A = \begin{pmatrix} 0 & -1 \\ 1 & 0 \end{pmatrix}$$

This is rotation by $\pi/2$ radians. Computing the characteristic polynomial as above we find that

$$p(t) = t^2 + 1$$

which has no real roots. Not surprisingly, A has no real eigenvalues.

(3) Finally, consider the real 3×3 matrix

$$A = \begin{pmatrix} 5 & 1 & 2 \\ 1 & 6 & 1 \\ 2 & 3 & 7 \end{pmatrix}$$

One may verify by direct calculation that 4 is a root of $\det(tI_3 - A)$, the characteristic polynomial of A, and is therefore an eigenvalue of A. To find the complete corresponding eigenspace, we solve the homogeneous linear system

$$(4I_3 - A)x = 0$$

which by construction must admit nontrivial solutions. We can equally well solve

$$(A - 4I_3)x = 0$$

which, in this case, is a trifle easier to handle. This amounts to the singular system

$$\begin{pmatrix} 1 & 1 & 2 \\ 1 & 2 & 1 \\ 2 & 3 & 3 \end{pmatrix} \begin{pmatrix} x_1 \\ x_2 \\ x_3 \end{pmatrix} = \begin{pmatrix} 0 \\ 0 \\ 0 \end{pmatrix}$$

The student may readily verify that the general solution is $\{(-3z,z,z) : z \in \mathbf{R}\}$, so that the eigenspace belonging to 4 is one-dimensional and spanned by the particular eigenvector $(-3,1,1)$.

Theorem 9-3 gives us, in principle at least, a way to determine the eigenvalues of a matrix considered as a linear transformation on k^n. In the more general case of an endomorphism T of a finite-dimensional vector space V, we can represent T by a matrix A relative to a given basis and then proceed as above. Both A and T will manifest exactly the same eigenvalues. (This is clear from the commutative diagram which defines the matrix of a linear transformation relative to a basis; see Section 6.3.) While A depends on the choice of basis, the characteristic polynomial does not. This follows at once from our next proposition:

9-4 PROPOSITION. *If two matrices are similar, then they have the same characteristic polynomial.*

Thus if $T : V \to V$ is any endomorphism of a finite-dimensional vector space, we may speak of the *characteristic polynomial of* T as the characteristic polynomial of *any* matrix representing T. Since by Theorem 6-23 all such representations are similar, this polynomial is independent of the choice of basis.

PROOF. Suppose that $B = P^{-1}AP$. Then by the rules of matrix arithmetic (especially the distributive law),

$$tI_n - B = tI_n - P^{-1}AP$$
$$= P^{-1}(tI_n)P - P^{-1}AP$$
$$= P^{-1}(tI_n - A)P$$

Here we have also used the property that scalar matrices (i.e., scalar multiples of the identity matrix) commute with all matrices of the same size. From this equation and the multiplicativity of the determinant, we find that

$$\det(tI_n - B) = \det(P^{-1})\det(tI_n - A)\det(P)$$
$$= \det(P)^{-1}\det(tI_n - A)\det(P)$$
$$= \det(tI_n - A)$$

so that by definition both matrices have the same characteristic polynomial. $\square$

We now explore some basic structural properties of eigenvalues and eigenvectors.

9-5 PROPOSITION. *Let $\lambda_1,\ldots,\lambda_r$ be distinct eigenvalues of $T:V \to V$ with corresponding nonzero eigenvectors $v_1,\ldots,v_r$. Then $v_1,\ldots,v_r$ are linearly independent.*

PROOF. Suppose that $v_1,\ldots,v_r$ are linearly dependent. Then there is a shortest possible dependence relation among the v_j, and we assume that this involves s of these vectors ($s>1$). After renumbering (if necessary), we have a relation of the form

$$v_1 = \mu_2 v_2 + \cdots + \mu_s v_s$$

where none of the coefficients $\mu_j \in k$ is zero. (Otherwise, we have a shorter dependence relation.) Apply T to both sides of the equation to obtain

$$\lambda_1 v_1 = \mu_2 \lambda_2 v_2 + \cdots + \mu_s \lambda_s v_s$$

Now subtract λ_1 times our first equation from the second. This yields

$$0 = \mu_2(\lambda_2 - \lambda_1)v_2 + \cdots + \mu_s(\lambda_s - \lambda_1)v_s$$

But since the λ_j are distinct and the μ_j are nonzero, this is an even shorter dependence relation among the v_j, and, as such, a clear contradiction. Thus the v_j must indeed be linearly independent, as claimed. $\square$

This leads at once to a beautiful result on diagonalization.

9-6 THEOREM. *Let $A \in M_n(k)$ and assume that the characteristic polynomial of A has n distinct roots in k. Then A is diagonalizable.*

PROOF. The n distinct roots of the characteristic polynomial are the eigenvalues of A. By the previous result, the corresponding eigenvectors constitute a linearly independent set of n vectors in k^n and hence are a basis. Since k^n admits an eigenbasis with respect to A, A is diagonalizable by Theorem 9-1.   □

9.2 Hermitian and Unitary Transformations

This section analyzes two special classes of endomorphisms on inner product spaces, the first of which includes symmetric matrices. As far as possible, we give a unified treatment of both real and complex spaces. Thus *inner product space* means either real or complex inner product space unless otherwise noted.

DEFINITION. Let $T: V \to V$ be an endomorphism of an inner product space V. Then an endomorphism T^* defined on the same space is called an *adjoint* for T if it satisfies the following condition:

$$\langle T(u)|v \rangle = \langle u|T^*(v) \rangle \quad \forall u,v \in V$$

T is called *self-adjoint* if $T^* = T$.

We shall show in the exercises that in the case of a finite-dimensional inner product space, adjoints exist and are unique. (See Exercise 18.) Hence in this case we may speak of *the* adjoint of an endomorphism. Note further the identity $T^{**} = T$. This is shown by the following elementary calculation, which holds for all vectors u and v in either a real or complex vector space:

$$\langle T(u)|v \rangle = \langle u|T^*(v) \rangle = \overline{\langle T^*(v)|u \rangle} = \overline{\langle v|T^{**}(u) \rangle} = \langle T^{**}(u)|v \rangle$$

The point is that since $T(u)$ and $T^{**}(u)$ have the same inner product with every vector, the difference $T(u) - T^{**}(u)$ is orthogonal to everything in V and hence must be the zero vector.

Let us now consider real and complex matrices. As a preliminary, if $A = (a_{ij})$ is a complex matrix, then its *conjugate* $\overline{A}$ is $(\overline{a}_{ij})$, the matrix obtained from A by replacing every entry by its conjugate. It is easy to show that

$$\overline{A+B} = \overline{A} + \overline{B} \quad \text{and} \quad \overline{AB} = \overline{A}\,\overline{B}$$

whenever these expressions are defined. If $A = (a_{ij})$, the *conjugate transpose* of A, denoted A^*, is appropriately enough the transpose of its conjugate; that is, the matrix whose (i,j)-entry is $\bar{a}_{ji}$. Note that for real matrices, $A^* = {}^tA$. Not surprisingly, $(AB)^* = B^*A^*$ whenever these products are defined.

9-7 LEMMA. *For all w, z in $\mathbf{C}^n$, $\langle Aw|z\rangle = \langle w|A^*z\rangle$. Hence A^* is the adjoint of A viewed as an endomorphism of $\mathbf{C}^n$. In particular, for real matrices A, the transpose is the adjoint.*

PROOF. This is an easy calculation using the elementary fact

$$\langle w|z\rangle = {}^tw\,\bar{z}$$

(the right side to be read as a matrix product) together with the other basic properties of transposition and conjugation. Thus

$$\langle Aw|z\rangle = {}^t(Aw)\bar{z} = {}^tw\,{}^tA\bar{z} = {}^tw\,\overline{A^*z} = \langle w|A^*z\rangle$$

as claimed.   □

We now introduce the first special class of endomorphisms to be considered here.

DEFINITION. An endomorphism T of a finite-dimensional inner product space is called *Hermitian* if it is self-adjoint; that is, if $T = T^*$. Accordingly, a matrix A is called *Hermitian* if $A = A^*$. (In the real case, this reduces to a symmetric matrix.)

It follows by definition that for a Hermitian mapping

$$\langle T(u)|v\rangle = \langle u|T(v)\rangle \quad \forall u,v \in V$$

Moreover, if $B = u_1, \ldots, u_n$ is an orthonormal basis for V, then T is Hermitian if and only if the matrix $A = (a_{ij})$ of T with respect to B is Hermitian. To see this, note that

$$\langle T(u_j)|u_i\rangle = \langle \sum_{k=1}^{n} a_{kj}u_k|u_i\rangle = \langle a_{ij}u_i|u_i\rangle = a_{ij}$$

while

$$\langle u_j|T(u_i)\rangle = \langle u_j|\sum_{k=1}^{n} a_{ki}u_k\rangle = \langle u_j|a_{ji}u_j\rangle = \bar{a}_{ji}$$

Clearly these two expressions are equal if and only if $A = A^*$.

This brings us to two fundamental, related properties of Hermitian transformations:

9-8 PROPOSITION. *All of the roots of the characteristic polynomial and hence all of the eigenvalues of a Hermitian transformation are real. Moreover, eigenvectors belonging to distinct eigenvalues are orthogonal.*

PROOF. Let $T: V \to V$ be Hermitian. Then as we have just seen, relative to an orthonormal basis for V, the matrix of T is also Hermitian, and so it suffices to analyze the eigenvalues of a Hermitian matrix A. One subtle but important point is that regardless of whether V is a real or complex inner product space, the matrix A operates on the complex space $\mathbf{C}^n$ by left multiplication. We need this extension because only over the complex numbers can we assume the existence of eigenvalues and eigenvectors for A.

Let λ be an eigenvalue for A with corresponding nonzero eigenvector w in $\mathbf{C}^n$. Then we have the following chain of equalities:

$$\lambda\langle w|w\rangle = \langle \lambda w|w\rangle = \langle A\,w|w\rangle = \langle w|A\,w\rangle = \langle w|\lambda w\rangle = \overline{\lambda}\langle w|w\rangle$$

Since w is not zero, neither is $\langle w|w\rangle$. We can thus cancel this factor from both sides to obtain

$$\lambda = \overline{\lambda}$$

This can only happen if $\lambda \in \mathbf{R}$, establishing the first statement. For the second, assume that λ_1 and λ_2 are distinct (necessarily real) eigenvalues for A, with corresponding eigenvectors w_1 and w_2. Then

$$\lambda_1\langle w_1|w_2\rangle = \langle \lambda_1 w_1|w_2\rangle = \langle A\,w_1|w_2\rangle = \langle w_1|A\,w_2\rangle = \langle w_1|\lambda_2 w_2\rangle = \lambda_2\langle w_1|w_2\rangle$$

and since $\lambda_1 \neq \lambda_2$, it follows that $\langle w_1|w_2\rangle = 0$, as claimed. $\square$

One aspect of this proof deserves additional comment. The mechanism by which we interpreted T as a linear transformation on a complex inner product space via its matrix A is somewhat artificial. A more abstract construction called the *tensor product*, which does not depend upon bases, is more natural, but beyond the scope of our discussion.

The next class of endomorphisms to be considered has two striking geometric properties that follow directly from a strictly algebraic definition.

DEFINITION. An invertible endomorphism T of a finite-dimensional inner product space is called a *unitary transformation* if its adjoint is equal to its inverse; that is, if $T^{-1} = T^*$. Accordingly, an invertible matrix A is called a *unitary matrix* if $A^{-1} = A^*$. A real unitary matrix is called an *orthogonal matrix*.

Recalling the definition of the adjoint, we have at once that for a unitary mapping

$$\langle T(u)|T(v)\rangle = \langle u|T^*T(v)\rangle = \langle u|v\rangle$$

so that unitary transformations preserve lengths and angles. The converse also holds; in fact, a weaker condition suffices.

9-9 PROPOSITION. *Suppose that T satisfies the condition*

$$\langle T(u)|T(u)\rangle = \langle u|u\rangle$$

for all $u \in V$, so that T preserves lengths. Then T is unitary.

PROOF. We expand both sides of the equality

$$\langle T(u+v)|T(u+v)\rangle = \langle u+v|u+v\rangle$$

to obtain

$$\langle T(u)|T(u)\rangle + \langle T(u)|T(v)\rangle + \langle T(v)|T(u)\rangle + \langle T(v)|T(v)\rangle = \langle u|u\rangle + \langle u|v\rangle + \langle v|u\rangle + \langle v|v\rangle$$

Canceling the outer terms from each side yields

$$\langle T(u)|T(v)\rangle + \langle T(v)|T(u)\rangle = \langle u|v\rangle + \langle v|u\rangle$$

and in the real case, this suffices. In the complex case, both sides exhibit a sum of conjugates. Thus the imaginary parts cancel, resulting in the equality

$$\mathrm{Re}(\langle T(u)|T(v)\rangle) = \mathrm{Re}(\langle u|v\rangle)$$

Substituting iu for u into this equation and exploiting the linearity of both T and the inner product, we find also that

$$\mathrm{Im}(\langle T(u)|T(v)\rangle) = \mathrm{Im}(\langle u|v\rangle)$$

since clearly $\mathrm{Re}(iz) = -\mathrm{Im}(z)$ for all complex numbers z. Hence both the real and imaginary parts of $\langle u|v\rangle$ and $\langle T(u)|T(v)\rangle$ are equal, and so these numbers must themselves be equal. This completes the proof.   □

Unitary matrices have an alternative characterization. Since the (i,j)-entry of $A\,{}^t\overline{A}$ is

the complex inner product of the ith row of A with the jth column of ${}^t A$

which is the same as

the complex inner product of the ith row of A with the jth row of A

it follows that $A\,{}^t\overline{A}$ equals $I_n = (\delta_{ij})$ if and only if the rows of A are orthonormal. Since the transpose of a unitary matrix is also unitary (exercise), the same result applies to the columns of A.

9-10 PROPOSITION. *The eigenvalues of a unitary transformation have absolute value 1. Moreover, eigenvectors belonging to distinct eigenvalues are orthogonal.*

PROOF. Let $T: V \to V$ be unitary and let λ be an eigenvalue, with corresponding nonzero eigenvector u. Then

$$\langle u|u \rangle = \langle T(u)|T(u) \rangle = \langle \lambda u | \lambda u \rangle = \lambda\overline{\lambda}\langle u|u \rangle$$

and it follows that

$$\lambda\overline{\lambda} = 1$$

establishing the first statement. For the second, assume that λ_1 and λ_2 are distinct eigenvalues with corresponding eigenvectors u_1 and u_2. Then

$$\langle u_1 | u_2 \rangle = \langle T(u_1)|T(u_2) \rangle = \langle \lambda_1 u_1 | \lambda_2 u_2 \rangle = \lambda_1\overline{\lambda_2}\langle u_1|u_2 \rangle = \frac{\lambda_1}{\lambda_2}\langle u_1|u_2 \rangle$$

(The last equality in the chain depends upon λ_2 having absolute value 1.) Now since by assumption $\lambda_1 \neq \lambda_2$, it follows that $\lambda_1/\lambda_2 \neq 1$. Therefore $\langle u_1|u_2 \rangle = 0$, as claimed. ◻

9.3 Spectral Decomposition

This brief section uses the basic properties of Hermitian and unitary transformations to prove one of the most beautiful and important results in mathematics. We begin with an easy technical lemma that isolates an essential common feature of both classes of transformations.

9-11 LEMMA. *Let T be either Hermitian or unitary. If u is an eigenvector of T and v is any vector orthogonal to u, then $T(v)$ remains orthogonal to u.*

PROOF. Suppose first that T is Hermitian and that u is an eigenvector belonging to the eigenvalue λ. Then

$$\lambda\langle u|v\rangle = \langle \lambda u|v\rangle = \langle T(u)|v\rangle = \langle u|T(v)\rangle$$

Hence if $\langle u|v\rangle=0$, also $\langle u|T(v)\rangle=0$, as claimed. Next suppose that T is unitary with u and λ as given previously. In this case, however, λ cannot be 0 since all of the eigenvalues of T have absolute value 1. (Alternatively, one could argue that T is invertible and hence has trivial kernel.) Then

$$\langle u|v\rangle = \langle T(u)|T(v)\rangle = \langle \lambda u|T(v)\rangle = \lambda\langle u|T(v)\rangle$$

Hence again $\langle u|v\rangle=0$ implies that $\langle u|T(v)\rangle=0$. This completes the proof.   □

If $T:V\to V$ is any endomorphism and W is a subspace of V, we say that W is *T-invariant* if T maps W into W; that is, if

$$T(w)\in W \quad \forall w\in W$$

Under these conditions it is clear that T restricts to an endomorphism $T|_W$ of W. The following observations are elementary:

(i)   The eigenvalues and eigenvectors of the restricted map $T|_W$ are also eigenvalues and eigenvectors of the original map T.

(ii) If T is Hermitian, then so is $T|_W$, since if $\langle T(u)|v\rangle=\langle u|T(v)\rangle$ in V, then certainly the same relation holds in W.

(iii) If T is unitary, then so is $T|_W$, since if $\langle T(u)|T(v)\rangle=\langle u|v\rangle$ in V, then certainly the same relation holds in W.

This brings us to our major result:

9-12 THEOREM. (Spectral Decomposition) *Let T be an endomorphism of the finite-dimensional inner product space V and suppose that either*

(i)   *T is Hermitian (in which case V may be defined either over the real or the complex numbers), or*

(ii) *T is unitary and V is defined over the complex numbers.*

Then there exists an orthonormal eigenbasis for V with respect to T. In particular, T is diagonalizable.

PROOF. The assumptions insure that all of the eigenvalues of T lie in the ground field for V: in the Hermitian case we know that the eigenvalues are real; in the unitary case, we work over the complex numbers, which are algebraically closed and therefore contain all of the roots of the characteristic polynomial of T. Assume that $\dim(V)=n$.

Let λ_1 be an eigenvalue of T and let u_1 be a corresponding eigenvector, which we normalize to unit length. Then u_1 spans a one-dimensional subspace W of V with orthogonal complement $W^\perp$ of dimension $n-1$. (Recall from Section 7.2 that $W^\perp$ is the subspace of all vectors orthogonal to W.) Evidently T maps W into itself, and, by the preceding lemma, T also maps $W^\perp$ into $W^\perp$. Thus $W^\perp$ is an invariant subspace for T. It follows from the observations preceding the theorem that T restricted to $W^\perp$ is also Hermitian or unitary, as the case may be, and that its eigenvalues and eigenvectors are also eigenvalues and eigenvectors for T. Thus iterating the process, or arguing by induction, we further obtain eigenvalues $\lambda_2,\ldots,\lambda_n$ (not necessarily distinct) and corresponding orthonormal eigenvectors $u_2,\ldots,u_n$. (The orthogonality follows by construction: at every step we choose u_j from a subspace orthogonal to all of its predecessors.) The collection $u_1,\ldots,u_n$ is thus the requisite orthonormal eigenbasis, and this completes the proof.   □

Exercises

1. Describe geometrically the linear transformation $T_A : \mathbf{R}^2 \to \mathbf{R}^2$ given in the first example in Section 9.1 and then interpret the meanings of the eigenvalues and eigenvectors accordingly.

2. Find one or more eigenvalue/eigenvector pairs other than exponentials of the form $Ce^{\lambda x}$ for the second derivative operator $D^2 : \mathscr{C}^\infty(\mathbf{R}) \to \mathscr{C}^\infty(\mathbf{R})$.

3. Show that if $T : V \to V$ is a linear transformation which is *not* injective, then 0 is an eigenvalue of T. (*Hint*: A non-injective linear transformation has nonzero kernel.)

4. Find the characteristic polynomial, eigenvalues, and corresponding eigenvectors for the matrix

$$A = \begin{pmatrix} 1 & 2 \\ 0 & 4 \end{pmatrix}$$

5. For which real values of a does the matrix

$$A = \begin{pmatrix} 2 & a \\ -1 & 1 \end{pmatrix}$$

have real eigenvalues? State your answer as an inequality. (*Hint*: The characteristic polynomial is quadratic; you must insure that its discriminant is nonnegative.)

6.   Compute the characteristic polynomial and eigenvalues for the matrix

$$A = \begin{pmatrix} 1 & 0 & 0 \\ 0 & 4 & 2 \\ 0 & 2 & 1 \end{pmatrix}$$

7.   Show that the following matrix has no real eigenvalues. Interpret this geometrically.

$$A = \begin{pmatrix} 0 & 1 \\ -1 & 0 \end{pmatrix}$$

8.   Let $A \in M_n(\mathbf{R})$ where n is odd. Show that A has at least one real eigenvalue. (*Hint*: The degree of the characteristic polynomial is odd. What does this imply about its behavior at $\pm\infty$? Now recall the Intermediate Value Theorem from elementary calculus.)

9.   For any $A \in M_n(\mathbf{R})$, show that the number of complex roots of the characteristic polynomial is even. This gives an alternate approach to the previous exercise. [*Hint*: Show that if λ is a root, then so is its complex conjugate. Thus the complex (but not real) roots always occur in pairs, and their number must be even.]

10. Let $\lambda \in k$ be an eigenvalue of $A \in M_n(k)$. Show that λ^r is an eigenvalue of A^r, the rth power of A, $r \geq 0$.

11. Let $A \in M_n(k)$ be such that A^r (in this case, A to the power r, not the rth column of A) is the zero matrix for some $r \geq 1$. Show that all of A's eigenvalues are 0.

12. Show that the eigenvalues of a triangular matrix are precisely the diagonal entries.

13. Find an example of a real 2×2 matrix A which is not diagonalizable as an endomorphism of $\mathbf{R}^2$, but *is* diagonalizable as an endomorphism of $\mathbf{C}^2$.

14. Find all of the eigenvalues of the matrix

$$A = \begin{pmatrix} 0 & 2 \\ -2 & 4 \end{pmatrix}$$

15. Show that the matrix of the previous problem is *not* diagonalizable over either the real or complex numbers. (*Hint*: Find the eigenvectors by solving a small linear system.)

16. Find the eigenvalues of the matrix

$$A = \begin{pmatrix} 2 & 1 \\ 0 & 5 \end{pmatrix}$$

17. For the matrix A of the previous problem, find an invertible matrix P such that $P^{-1}AP$ is a diagonal matrix. (*Hint*: Find the eigenvectors corresponding to the eigenvalues of A; now recall the change of basis formula.)

18. Let V be a finite-dimensional inner product space (either real or complex) and let T be an endomorphism of V. In this problem we show that an adjoint endomorphism T^* exists and is unique. (Our approach assumes familiarity with Section 6.4.)

(a) Let $v \in V$. Define a mapping L_v from V to the ground field k by

$$L_v(u) = \langle u|v \rangle$$

Show that L_v is linear for all v, so that we have a mapping

$$L:V \to V^*$$
$$v \mapsto L_v$$

where $V^* = \text{Hom}(V,k)$ is the dual space.

(b) Show that L is linear in the case of a real inner product space and *antilinear* in the case of a complex inner product space. That is, in both cases L satisfies the identity

$$L_{v+w} = L_v + L_w$$

For real spaces we have also that

$$L_{\lambda v} = \lambda L_v$$

but for complex spaces this relation must be replaced by

$$L_{\lambda v} = \overline{\lambda} L_v$$

(c) Show that $L: V \to V^*$ is bijective, so that every element of the dual space may be expressed uniquely as L_v for some $v \in V$. (Take care to account for the not-quite linearity of L in the complex case.)

(d) Let T be any endomorphism of V. Fixing an element $v \in V$, show that the mapping $u \mapsto \langle T(u)|v \rangle$ lies in V^*. Observe now that according to part (c), there exists a unique element in V, which we shall denote $T^*(v)$, such that

$$\langle T(u)|v \rangle = \langle u|T^*(v) \rangle$$

for all $u \in V$.

(e) Show that the association $v \mapsto T^*(v)$ is linear, so that T^* is likewise an endomorphism of T. This is the long sought after adjoint map. Note that its uniqueness is implicit in our method of construction.

REMARK. One might alternatively make a direct computational argument based on the existence of an orthonormal basis. This is perhaps simpler, but certainly less instructive. The approach taken here also explains (if you look closely and compare to Section 6.4) why the notation T^* is used for both the adjoint map and the transpose map.

19. Let $A \in M_n(\mathbf{C})$. Show that (i) AA^* is Hermitian, and that (ii) $A^{**}=A$.

20. Let $A \in GL_n(\mathbf{C})$. Show that $(A^*)^{-1}=(A^{-1})^*$.

21. Find all of the eigenvalues of the matrix

$$A = \begin{pmatrix} 2 & 0 & 0 \\ 0 & 3 & 1 \\ 0 & 1 & 3 \end{pmatrix}$$

How does one know in advance that these will be real?

22. Find an orthonormal eigenbasis for $\mathbf{R}^3$ with respect to the matrix of the previous problem. What is the matrix of the given transformation with respect to this new basis? Verify the change of basis formula in this case.

23. Let V be a complex inner product space and let W be a finite-dimensional subspace of V. As usual, pr_W denotes the orthogonal projection map from V to W. Show that pr_W (considered as an endomorphism of V) is a Hermitian transform; that is, prove that for all x and y in V, $\langle \mathrm{pr}_W(x)|y \rangle = \langle x|\mathrm{pr}_W(y) \rangle$.

(*Hint*: Choose an orthonormal basis $u_1,...,u_m$ for W and directly compute both expressions.)

24. The projection map defined in the previous problem can have at most two eigenvalues. What are they? (*Hint*: Don't compute; conceptualize!)

25. Consider the differential operator $\mathscr{C}^\infty(\mathbf{R}) \to \mathscr{C}^\infty(\mathbf{R})$ defined by

$$D(y) = y'' + py' + qy$$

where p and q are real constants. Show that $e^{\lambda x}$ lies in the kernel of D if and only if λ is a root of the polynomial

$$p(t) = t^2 + pt + q$$

This is called the *characteristic polynomial* of the homogeneous equation $y'' + py' + qy = 0$.

26. Find two linearly independent solutions to the homogeneous differential equation

$$y'' - 5y' + 4y = 0$$

27. Find two linearly independent solutions to the homogeneous differential equation

$$y'' - 4y' + 4y = 0$$

This is harder than the previous problem. Why?

28. Generalize the result of Exercise 25 to the case of an nth-order linear differential operator

$$D(y) = \sum_{j=0}^{n} a_j \frac{d^j y}{dx^j} \quad (a_j \in \mathbf{R})$$

29. Let $A \in M_n(k)$ and let q be a polynomial with coefficients in k. Then since $M_n(k)$ is a k-algebra, it makes sense to consider the matrix $q(A)$, the result of evaluating q at A. [For instance, if $q(t) = t^2 + 5$, then $q(A) = A^2 + 5I_n$.] With this in mind, prove the following result:

If λ is an eigenvalue of A with corresponding nonzero eigenvector x, then $q(A)x = q(\lambda)x$ and hence $q(\lambda)$ is an eigenvalue of the matrix $q(A)$.

30. Let $A \in M_n(k)$ with characteristic polynomial $p(t)$ and suppose that $x \in k^n$ is an eigenvector of A. Using the previous problem, show that $p(A)x = 0$. [Actually, much more is true: $p(A)$ is in fact the zero matrix! This result, known as the Cayley-Hamilton Theorem, is proven in the next chapter.]

10
Triangulation and Decomposition of Endomorphisms

In Section 9.3 we saw that Hermitian and complex unitary transformations always admit an orthogonal basis of eigenvectors, with respect to which these mappings are represented by simple diagonal matrices. This chapter presses the attack to find out what in general can be said about an arbitrary endomorphism T of a finite-dimensional vector space over an abstract field k. We first establish an astounding property of the characteristic polynomial; this is the content of the Cayley-Hamilton Theorem (10-1). Next, using similar techniques, we show that T is representable at least by an upper triangular matrix, provided that the roots of its characteristic polynomial all lie in k. This leads to the introduction of so-called *nilpotent* mappings. Maintaining our previous assumption on the characteristic polynomial, we then show that T is expressible as the sum of a diagonal map (easily built out of its eigenvalues) and a nilpotent map. Finally, further analysis of nilpotent endomorphisms yields a special matrix representation called the Jordan normal form.

This material is comparatively difficult, although the principal results are easily understood. Almost all of the examples of the theory are given in the exercises.

Throughout this chapter all vector spaces are finite dimensional.

10.1 The Cayley-Hamilton Theorem

Let k be an arbitrary field, and let V be a vector space over k. Recall from Chapter 6 that $\mathrm{Hom}(V,V)$, the set of endomorphisms of V, constitutes a k-algebra. This is to say, in particular, that $\mathrm{Hom}(V,V)$ enjoys the following algebraic properties:

(i) With respect to addition and scalar multiplication of functions, $\mathrm{Hom}(V,V)$ is a vector space over k.

(ii) With respect to addition and composition of functions, $\mathrm{Hom}(V,V)$ is a ring with unity (cf. Section 2.3).

(iii) Scalar multiplication commutes with composition in the following sense:

$$(\lambda T_1)T_2 = T_1(\lambda T_2) = \lambda(T_1 T_2) \quad \forall \lambda \in k; T_1, T_2 \in \operatorname{Hom}(V,V)$$

(In this chapter we usually omit the composition operator and simply write $T_1 T_2$ for $T_1 \circ T_2$.)

Let $k[t]$ denote the vector space consisting of all polynomials in the indeterminate t with coefficients in the field k. This is also a k-algebra, in this case with respect to ordinary addition and multiplication of polynomials, and is always commutative. Let $p(t)$ be a polynomial in $k[t]$, so that $p(t)$ takes the form

$$p(t) = \sum_{j=0}^{r} \alpha_j t^j \quad (\alpha_0, \ldots, \alpha_r \in k)$$

Then given $T \in \operatorname{Hom}(V,V)$ it makes sense to evaluate $p(T)$:

$$p(T) = \sum_{j=0}^{r} \alpha_j T^j$$
$$= \alpha_r T^r + \alpha_{r-1} T^{r-1} + \cdots + \alpha_1 T + \alpha_0 1_V$$

Here T^j is the composition of T with itself j times, with the natural convention that T^0 is the identity map. Fixing T we can regard this as a mapping

$$k[t] \to \operatorname{Hom}(V,V)$$
$$p \mapsto p(T)$$

This map is obviously k-linear, but even more is true. Since powers of T commute with each other, one shows easily that for all polynomials $p, q \in k[t]$,

$$p(T)q(T) = pq(T)$$

where the right-hand side indicates that we first multiply p and q as polynomials in t and then evaluate the result at T. Thus evaluation at T is in fact a homomorphism of k-algebras.

Everything we have just discussed applies equally well to the matrix algebra $M_n(k)$ $(n \geq 1)$. In particular, if $p(t) \in k[t]$ and $A \in M_n(k)$, then it makes sense to evaluate $p(A)$:

$$p(A) = \sum_{j=0}^{r} \alpha_j A^j$$
$$= \alpha_r A^r + \alpha_{r-1} A^{r-1} + \cdots + \alpha_1 A + \alpha_0 I_n$$

With these ideas in mind, we can now state the main result of this section:

10-1 THEOREM. (Cayley-Hamilton) *Let V be a vector space of dimension $n \geq 1$ over the field k. If $T \in \text{Hom}(V,V)$ has characteristic polynomial $p(t)$, then $p(T)=0$; that is, $p(T)$ is the zero map on V. Equivalently, if $A \in M_n(k)$ has characteristic polynomial $p(t)$, then $p(A)=0$; that is, an $n \times n$ matrix always satisfies its characteristic polynomial.*

This theorem is nothing less than amazing, although not especially deep. We postpone its proof to make way for an example and for some preliminary analysis.

EXAMPLE

Recall that the matrix of counterclockwise rotation on $\mathbf{R}^2$ by $\pi/2$ radians about the origin is

$$A = \begin{pmatrix} 0 & -1 \\ 1 & 0 \end{pmatrix}$$

with characteristic polynomial $p(t) = t^2 + 1$. The Cayley-Hamilton Theorem asserts in this case that $A^2 + I_2$ is the 2×2 zero matrix. To verify this, we need only compute

$$A^2 = \begin{pmatrix} -1 & 0 \\ 0 & -1 \end{pmatrix}$$

so that indeed $A^2 + I_2 = 0$, as predicted.

The key to Theorem 10-1 is the analysis of the characteristic polynomial of an endomorphism satisfying a special condition identified in Lemma 10-2 below. To explain this, we revisit and highlight a definition made in passing near the end of the previous chapter—a definition that in fact generalizes the notion of an eigenspace.

DEFINITION. Let $T: V \rightarrow V$ be a linear transformation. Then a subspace W of V is called *invariant* with respect to T (or *T-invariant*) if $T(W) \subseteq W$; that is, if for all $w \in W$, $T(w) \in W$.

Thus W is T-invariant if and only if T restricts to an endomorphism of W. Clearly both V itself and the zero subspace are invariant subspaces for any endomorphism. The eigenspace W associated with any given eigenvalue λ of T is likewise T-invariant, since by construction $T(w) = \lambda w \in W$ for all $w \in W$.

10-2 LEMMA. *Let $T: V \to V$ be a linear transformation on a vector space V over k of dimension $n > 1$ and assume that T admits no proper, nontrivial invariant subspaces. Then there exists a basis B for V such that the matrix of T with respect to B takes the form*

$$A = \begin{pmatrix} 0 & 0 & 0 & \cdots & 0 & \alpha_0 \\ 1 & 0 & 0 & \cdots & 0 & \alpha_1 \\ 0 & 1 & 0 & \cdots & 0 & \alpha_2 \\ \vdots & \vdots & \vdots & & \vdots & \vdots \\ 0 & 0 & 0 & \cdots & 0 & \alpha_{n-2} \\ 0 & 0 & 0 & \cdots & 1 & \alpha_{n-1} \end{pmatrix}$$

for some scalars $\alpha_0, \ldots, \alpha_{n-1} \in k$. Moreover, $p(t)$, the characteristic polynomial of T is then given by

$$p(t) = t^n - \alpha_{n-1} t^{n-1} - \alpha_{n-2} t^{n-2} - \cdots - \alpha_1 t - \alpha_0$$

PROOF. Let v_0 be any nonzero vector in V. Now recursively define the elements $v_1, \ldots, v_{n-1}$ of V as follows:

$$v_{j+1} = T(v_j) \quad (j = 0, \ldots, n-2)$$

Equivalently, $v_j = T^j(v_0)$; we are simply following the trajectory of v_0 under successive applications of T. We argue by contradiction that the collection of vectors $v_0, \ldots, v_{n-1}$ is linearly independent and hence constitutes a basis B for V. Suppose otherwise. Then there is a smallest j such that $v_0, \ldots, v_j$ are linearly dependent. Clearly $0 < j \leq n-1$ and v_j must be involved in the dependence relation. By Lemma 4-2, v_j then lies in $W = \mathrm{Span}(v_0, \ldots, v_{j-1})$. But then $T(v_0), \ldots, T(v_{j-1})$ all lie in W, which is therefore a T-invariant subspace of V. Moreover, $v_0 \in W$, so W is nontrivial, and W is spanned by fewer than n vectors, so W is also proper. This contradicts our hypothesis that no such invariant subspace exists.

Given this basis B, the first part of the lemma follows immediately. For we know that

$$T(v_j) = v_{j+1} \quad (0 \leq j < n-1) \quad \text{and} \quad T(v_{n-1}) = \sum_{j=0}^{n-1} \alpha_j v_j$$

for some family of scalars $\alpha_j \in k$, whence the matrix of T relative to B is the matrix A, exactly as stated.

The second part of the lemma requires that we compute $p(t)$, the characteristic polynomial of A:

$$p(t) = \det \begin{pmatrix} t & 0 & 0 & \cdots & 0 & -\alpha_0 \\ -1 & t & 0 & \cdots & 0 & -\alpha_1 \\ 0 & -1 & t & \cdots & 0 & -\alpha_2 \\ \vdots & \vdots & \vdots & & \vdots & \vdots \\ 0 & 0 & 0 & \cdots & t & -\alpha_{n-2} \\ 0 & 0 & 0 & \cdots & -1 & t-\alpha_{n-1} \end{pmatrix}$$

We shall show inductively, starting at $n = 2$, that $p(t)$ reduces to the formula given above. For $n=2$ this is merely the calculation

$$\det \begin{pmatrix} t & -\alpha_0 \\ -1 & t-\alpha_1 \end{pmatrix} = t^2 - \alpha_1 t - \alpha_0$$

For $n > 2$, we expand by the first row. This yields two terms, corresponding to the first and last columns. The first term is

$$t \cdot \det \begin{pmatrix} t & 0 & 0 & \cdots & 0 & -\alpha_1 \\ -1 & t & 0 & \cdots & 0 & -\alpha_2 \\ 0 & -1 & t & \cdots & 0 & -\alpha_3 \\ \vdots & \vdots & \vdots & & \vdots & \vdots \\ 0 & 0 & 0 & \cdots & t & -\alpha_{n-2} \\ 0 & 0 & 0 & \cdots & -1 & t-\alpha_{n-1} \end{pmatrix}$$

which, by the induction hypothesis, amounts to

$$t(t^{n-1} - \alpha_{n-1}t^{n-2} - \cdots - \alpha_1) = t^n - \alpha_{n-1}t^{n-1} - \cdots - \alpha_1 t$$

Thus it remains to show that the second term in the expansion, which is given by

$$-\alpha_0(-1)^{n+1}\det\begin{pmatrix} -1 & t & 0 & \cdots & 0 \\ 0 & -1 & t & \cdots & 0 \\ 0 & 0 & -1 & \cdots & 0 \\ \vdots & \vdots & \vdots & & \vdots \\ 0 & 0 & 0 & \cdots & t \\ 0 & 0 & 0 & \cdots & -1 \end{pmatrix}$$

(remember the alternating signs in the expansion rule) is $-\alpha_0$. But this is wonderful! The matrix in question is diagonal, whence its determinant is just the product of the $n-1$ diagonal entries. Now watch how the signs balance:

$$-\alpha_0(-1)^{n+1}(-1)^{n-1} = -\alpha_0(-1)^{2n} = -\alpha_0$$

Thus the second term is indeed $-\alpha_0$, completing the proof of the lemma.   ❑

Note that the second statement of the lemma is also true in the one-dimensional case. For then the endomorphism T is scalar multiplication by some α_0 and the characteristic polynomial is just $T-\alpha_0$.

This brings us to the heart of the Cayley-Hamilton Theorem, here stated as our final lemma.

10-3 LEMMA. *Let $T:V \rightarrow V$ be a linear transformation on a vector space V over k of dimension $n \geq 1$ and assume that T admits no proper, nontrivial invariant subspaces. Let $p(t)$ denote the characteristic polynomial of T. Then $p(T)=0$.*

PROOF. Let v_j and α_j ($j=0,\dots,n-1$) be as in the previous lemma and its proof. Recall in particular that

$$v_j = T^j(v_0) \ (0 \leq j < n-1) \quad \text{and} \quad T(v_{n-1}) = \sum_{j=0}^{n-1}\alpha_j v_j$$

Using Lemma 10-2, we first compute the effect of $p(T)$ on the basis vector v_0:

$$p(T)(v_0) = T^n(v_0) - \alpha_{n-1}T^{n-1}(v_0) - \cdots - \alpha_1 T(v_0) - \alpha_0 v_0$$
$$= T(v_{n-1}) - \alpha_{n-1}v_{n-1} - \cdots - \alpha_1 v_1 - \alpha_0 v_0$$
$$= 0$$

Next consider the effect of $p(T)$ on any other v_j ($j>0$):

$$p(T)(v_j) = p(T)(T^j(v_0))$$
$$= [p(T)T^j](v_0)$$
$$= [T^j p(T)](v_0)$$
$$= T^j(p(T)(v_0))$$
$$= T^j(0)$$
$$= 0$$

The key point is that in the algebra $\text{Hom}(V,V)$, the maps $p(T)$ and T^j commute, since they involve only powers of T and scalars. Hence knowing that $p(T)$ sends v_0 to 0, we infer that $p(T)$ also sends every element of the basis $v_0,\ldots,v_{n-1}$ to 0. Therefore $p(T)$ is itself the zero map, as claimed.   □

We now have all of the tools needed to prove Theorem 10-1.

Proof of Theorem 10-1

We work by induction on n, the dimension of V over k. The case $n=1$ is covered by the preceding lemma, since a vector space of dimension 1 evidently has no proper, nontrivial subspaces whatsoever. Assume now that $n>1$. Then if V has no proper, nontrivial T-invariant subspaces, we may again appeal to Lemma 10-3 to conclude that T satisfies $p(t)$, the characteristic polynomial of T. Thus we may proceed under the additional hypothesis that V does admit a T-invariant subspace W_1 which is neither all of V nor the zero subspace.

Let B_1 be a basis for W_1. Then we can extend B_1 to a basis B for all of V by adjoining a linearly independent collection B_2 to B_1. These additional vectors in B_2 themselves span a subspace W_2 of V (also both proper and nontrivial), and in fact $W = W_1 \oplus W_2$. Now consider A, the matrix of T with respect to B. Since T sends W_1 into itself, A has the following block decomposition:

$$A = \left(\begin{array}{c|c} P & Q \\ \hline 0 & R \end{array} \right)$$

where P and R are square matrices whose sizes are, respectively, $\dim(W_1)$ and $\dim(W_2)$. Note that W_2 is not necessarily T-invariant, so the remaining block Q does not necessarily consist entirely of zeros. Let T_R be the linear transformation from W_2 to itself whose matrix is R relative to the basis B_2. We need to investigate a special relationship between T and T_R, but first observe that by Corollary 8-8 the polynomial $p(t)$ has factorization

$$p(t) = p_1(t)p_2(t)$$

where $p_1(t)$ is the characteristic polynomial of T restricted to an endomorphism of W_1 (i.e., the characteristic polynomial of the matrix P) and $p_2(t)$ is the characteristic polynomial of T_R (i.e., the characteristic polynomial of the matrix R).

From the structure of A, we deduce at once that given $w \in W_2$, $T(w)$ and $T_R(w)$ differ by an element of W_1; that is,

$$T(w) - T_R(w) \in W_1$$

for all $w \in W_2$. (The point is that T_R acting on an element of B_2 yields the linear combination of elements in B_2 specified by the matrix R; T acting on the same element yields the same linear combination of elements in B_2, but then adds the linear combination of elements in B_1 specified by the matrix Q.) Applying T again and recalling the invariance of W_1, we find further that

$$T^2(w) - TT_R(w) \in W_1$$

for all $w \in W_2$. Since replacing T by T_R in the second term only costs us another element of W_1, we conclude moreover that

$$T^2(w) - T_R^2(w) \in W_1$$

for all $w \in W_2$. Continuing in this way, we see that for all nonnegative integers j

$$T^j(w) - T_R^j(w) \in W_1 \quad \forall w \in W_2$$

Consequently, for any polynomial $q(t) \in k[t]$, we have

$$q(T)(w) - q(T_R)(w) \in W_1 \quad \forall w \in W_2 \tag{10.1}$$

So then, consider the endomorphism $p(T) = p_1(T)p_2(T) = p_2(T)p_1(T)$. What is its effect on elements of W_1? By induction, $p_1(T)$ is the zero map on W_1, so for any $w \in W_1$,

$$p(T)(w) = [p_2(T)p_1(T)](w)$$
$$= p_2(T)(0)$$
$$= 0$$

And what is the effect of $p(T)$ on elements of W_2? In this case, we note that according to our general analysis (Eq. 10.1),

$$p_2(T)(w) - p_2(T_R)(w) \in W_1$$

for $w \in W_2$. But again invoking the induction hypothesis, $p_2(T_R)$ is the zero map on W_2, from which we deduce at once that $p_2(T)(w) \in W_1$ for all $w \in W_2$. Therefore, given any $w \in W_2$, $p_2(T)(w) = w_1$ for some $w_1 \in W_1$ and

$$
\begin{aligned}
p(T)(w) &= [p_1(T)p_2(T)](w) \\
&= p_1(T)(w_1) \\
&= 0
\end{aligned}
$$

Thus $p(T)$ is the zero map on both W_1 and W_2 and hence on all of V. This completes the proof. $\square$

10.2   Triangulation of Endomorphisms

Let k be a field. Recall that a matrix $A = (a_{ij}) \in M_n(k)$ is called *upper triangular* if $a_{ij} = 0$ whenever $j < i$; that is, if A has the form

$$
A = \begin{pmatrix}
a_{11} & a_{12} & a_{13} & \cdots & a_{1n} \\
0 & a_{22} & a_{23} & \cdots & a_{2n} \\
0 & 0 & a_{33} & \cdots & a_{3n} \\
\vdots & \vdots & \vdots & & \vdots \\
0 & 0 & 0 & \cdots & a_{nn}
\end{pmatrix}
$$

Evidently the upper triangular matrices constitute a subring of $M_n(k)$, and in fact the invertible upper triangular matrices constitute a subgroup of $GL_n(k)$. (See Exercise 5 below.) Since we shall have no occasion to refer to lower triangular matrices here, we make the following convention:

Henceforth in this chapter, the phrase "triangular matrix" means "upper triangular matrix."

Let T be an endomorphism of the vector space V. Then if we can find a basis B for V such that the matrix of T with respect to B is triangular, we say that T can be *reduced to triangular form*. One advantage of such a reduction is that the eigenvalues of T are then apparent; another is that nonsingular triangular matrices are easily inverted (Exercise 6). We now state and prove a theorem giving conditions under which such a reduction is possible.

10-4 THEOREM. *Let V be a vector space of dimension n over the field k, and let T be an endomorphism of V. Suppose that the characteristic polynomial of T has all of its roots in k. Then T may be reduced to triangular form.*

REMARKS. The meaning of the statement that a polynomial $p(t) \in k[t]$ has all of its roots in k is clear enough when k is the field of real or complex numbers, but perhaps not when k is an abstract field. In this more general case, it simply means that there is no larger field K containing k in which new roots of $p(t)$ appear. This is equivalent to the assertion that $p(t)$ admits a linear factorization in $k[t]$; that is, there exist elements $c, \alpha_1, \ldots, \alpha_n \in k$ such that

$$p(t) = c \prod_{j=1}^{n} (t - \alpha_j)$$

In one direction, the implication is clear: if $p(t)$ admits such a factorization, its only possible roots are $\alpha_1, \ldots, \alpha_n$, which by assumption all lie in k. The other direction is far more delicate and depends upon the fact that a nonconstant polynomial in $k[t]$ always has a root in some extension field of k.

PROOF. The proof, which is by induction on n, leans heavily on some of the ideas of the proof of the Cayley-Hamilton Theorem. (The common thread here, as the student may see later in a course in abstract algebra, is that the induction step proceeds essentially via the *quotient space*, a notion beyond the scope of this text.) The case $n = 1$ is trivial (every 1×1 matrix is triangular!), so we assume that $n > 1$.

Let $p(t)$ be the characteristic polynomial of T, and let λ_1 be a root of $p(t)$. By hypothesis and by the remarks above, such a root exists and lies in k. Thus λ_1 is an eigenvalue for T. Let v_1 be a corresponding nonzero eigenvector. Then v_1 serves as basis for a one-dimensional T-invariant subspace W_1 of V. Now extend v_1 to a basis $B = \{v_1, v_2, \ldots, v_n\}$ for all of V, and let W_2 be the subspace spanned by $v_2, \ldots, v_n$. Accordingly, we have $V = W_1 \oplus W_2$ (although W_2 is not necessarily also T-invariant). The matrix of T with respect to B has the form

$$A = \left(\begin{array}{c|c} \lambda_1 & * \\ \hline 0 & R \end{array} \right)$$

where R is an $(n-1) \times (n-1)$ matrix and the asterisk denotes a row of scalars, in which we shall have no further interest. Let T_R denote the endomorphism of W_2 represented by R with respect to the basis $v_2, \ldots, v_n$. Then evidently T and T_R acting on W_2 only differ by elements of W_1—a critical point in the sequel.

Next consider $q(t)$, the characteristic polynomial of T_R. Expanding the determinant of $tI_n - A$ by the first column shows at once that

$$p(t) = (t - \lambda_1)q(t)$$

whence every root of $q(t)$ is likewise a root of $p(t)$. It follows then that all of the roots of $q(t)$ also lie in k. Hence by induction, there exists a basis $v_2', \ldots, v_n'$ for

W_2 with respect to which the matrix of T_R is triangular. Call this new matrix R'. Now the collection of vectors $v_1, v_2', \ldots, v_n'$ constitutes a basis B' for V, and since T and T_R only differ by elements of W_1 on each of the vectors $v_2', \ldots, v_n'$, it follows that the matrix of T with respect to B' has the form

$$A' = \left(\begin{array}{c|c} \lambda_1 & * \\ \hline 0 & R' \end{array} \right)$$

But A' is manifestly triangular because R' is, and this completes the proof.   ☐

Our first corollary is just the interpretation of this theorem for matrices via the change of basis formula (Theorem 6-23).

10-5 COROLLARY. *Let $A \in M_n(k)$, and suppose that all of the roots of the characteristic polynomial of A lie in k. Then there exists a matrix P in $GL_n(k)$ such that $B = P^{-1}AP$ is triangular.*   ☐

Since the complex numbers are algebraically closed (that is, every polynomial with complex coefficients has all of its roots in the complex numbers), Theorem 10-4 also yields the following:

10-6 COROLLARY. *Every endomorphism of a complex vector space may be reduced to triangular form.*   ☐

This result holds, of course, for *any* algebraically closed field. (There are others besides **C**; the interested student should hurry on to an abstract algebra course.)

The next result adds some extra information to Theorem 10-4 in the special case of an inner product space. Note that it does *not* hold (or even make sense) over arbitrary fields.

10-7 THEOREM. *Let V be a real or complex inner product space and let T be an endomorphism of V. If V is defined over **R**, assume further that all of the roots of the characteristic polynomial of T are real. Then there exists an orthonormal basis for V with respect to which the matrix of T is triangular.*

PROOF. The proof is identical to that of Theorem 10-4 except that in the induction step we choose $W_2 = W_1^\perp$, the orthogonal complement of W_1. We leave the details to the reader.   ☐

The final result of the section is just the interpretation of this result for real or complex matrices.

10-8 COROLLARY. *Let A be a real or complex square matrix. In the case that A is real, assume further that all of the roots of the characteristic polynomial of A are likewise real. Then there exists a unitary matrix P such that $B = P^{-1}AP$ is triangular.*   ❏

The point is that the transition matrix P from an orthonormal basis of either $\mathbf{R}^n$ or $\mathbf{C}^n$ to the canonical basis is obviously unitary. (See the discussion following the definition of a unitary transformation in Section 9.2.)

10.3 Decomposition by Characteristic Subspaces

Throughout this section, V is a vector space over k of dimension n, and T is an endomorphism of V, with characteristic polynomial $p(t)$. Moreover, we assume that all of the roots of $p(t)$ lie in k.

For the discussion at hand we require the following terminology and results from elementary abstract algebra. (The proofs can be found in almost any book on the subject.)

A nonconstant polynomial $g(t) \in k[t]$ is called *irreducible* in $k[t]$ if for every factorization

$$g(t) = h_1(t)h_2(t) \quad (h_1(t), h_2(t) \in k[t])$$

either $h_1(t)$ or $h_2(t)$ is a constant polynomial. Clearly every nonzero linear polynomial (i.e., polynomial of first degree) is irreducible. (Note that irreducibility depends on the field k. See Exercise 8.)

FACT 1. (Unique Factorization for Polynomials) *Every nonconstant polynomial in $k[t]$ can be factored into the product of irreducible polynomials in $k[t]$. Moreover, apart from order and constant factors, this factorization is unique.*

FACT 2. *Let $g_1, \ldots, g_m$ be polynomials in $k[t]$ and assume that $g_1, \ldots, g_m$ have no nonconstant common factors. Then there exist polynomials $h_1, \ldots, h_m$ in $k[t]$ such that*

$$g_1(t)h_1(t) + \cdots + g_m(t)h_m(t) = 1$$

Note that the right side represents the constant polynomial 1.

We can now proceed with the main exposition. The assumptions laid out at the head of the section guarantee the existence of $\lambda_1, \ldots, \lambda_n \in k$ such that

$$p(t) = \prod_{j=1}^{n}(t - \lambda_j)$$

Of course, not all of the λ_j need be distinct. Suppose that r of them are. Then we may refine the factorization above to

$$p(t) = \prod_{j=1}^{r}(t - \lambda_j)^{m_j} \tag{10.2}$$

where the positive integer m_j is called the *multiplicity of the root* λ_j ($j = 1,...,r$). Define r corresponding endomorphisms $N_1,...,N_r$ of V by

$$N_j = (T - \lambda_j I_n)^{m_j}$$

and corresponding subspaces $U_1,...,U_r$ of V by

$$U_j = \text{Ker}(N_j)$$

($j = 1,...,r$). Observe that each U_j contains the eigenspace corresponding to λ_j, and hence is nontrivial. This brings us to a fundamental definition.

DEFINITION. The subspaces $U_1,...,U_r$ of V are called the *characteristic subspaces* of V with respect to the endomorphism T.

With these preliminaries in hand, we can now give a lovely generalization of the Spectral Decomposition Theorem (and also of Theorem 9-6). Note that we are restricted neither to Hermitian or unitary transformations nor to the field of real or complex numbers. The key assumption in force remains that all of the roots of the characteristic polynomial of T lie in the ground field k.

10-9 THEOREM. *Let* $U_1,...,U_r$ *be the characteristic subspaces of* V *with respect to* T. *Then with respect to the notation above, the following assertions hold:*

(i) *Each characteristic subspace* U_j ($j = 1,...,r$) *is T-invariant.*

(ii) $V = U_1 \oplus \cdots \oplus U_r$; *that is, each vector* $v \in V$ *can be expressed uniquely as a sum* $v = u_1 + \cdots + u_r$, *where* $u_j \in U_j$ ($j = 1,...,r$).

(iii) *T restricted to* U_j *has only one eigenvalue, and this is* λ_j ($j = 1,...,r$).

(iv) $\text{Dim}(U_j) = m_j$, *the multiplicity of* λ_j ($j = 1,...,r$).

PROOF. (i) With $N_1,...,N_r$ defined as above, we note that each N_j clearly commutes with T. By construction, $N_j(U_j)=\{0\}$, and therefore

$$N_j T(U_j)=TN_j(U_j)=T(\{0\})=\{0\}$$

showing that $T(U_j)$ lies in $\text{Ker}(N_j)=U_j$, as claimed.

(ii) According to Proposition 3-8, we must show that

(a) $V=U_1+\cdots+U_r$; that is, every element in V can be written as the sum of elements in $U_1,...,U_r$.

(b) If $u_1+\cdots+u_r=0$, where $u_j\in U_j$ $(j=1,...,r)$, then each u_j is itself zero.

We shall first demonstrate (a). Consider the following family of polynomials:

$$q_j(t)=\frac{p(t)}{(t-\lambda_j)^{m_j}} \quad (j=1,...,r)$$

All we have done here is to delete the factors corresponding to λ_j from $p(t)$. By construction and the Cayley-Hamilton Theorem,

$$[N_j q_j(T)](v)=p(T)(v)=0$$

and therefore $q_j(T)(v)\in U_j$ for all $v\in V$. Moreover, since the q_j have no nonconstant common factor, there exist polynomials $h_j(t)$ $(j=1,...,r)$ such that

$$h_1(t)q_1(t)+\cdots+h_r(t)q_r(t)=1$$

It follows that

$$h_1(T)q_1(T)+\cdots+h_r(T)q_r(T)=1_V \qquad (10.3)$$

Thus for any $v\in V$,

$$v=\sum_{j=1}^{r}[h_j(T)q_j(T)](v)$$

But since $q_j(T)(v)\in U_j$ and U_j is T-invariant, it follows that the jth summand in this expression does indeed lie in U_j $(j=1,...,r)$. This establishes point (a).

To prove (b), we first note a special property of the products $h_j(t)q_j(t)$ appearing in the previous paragraph. Given $u_k\in U_k$ $(k=1,...,r)$,

$$[h_j(T)q_j(T)](u_k) = \begin{cases} u_k & \text{if } j = k \\ 0 & \text{otherwise} \end{cases} \tag{10.4}$$

for $j = 1,...,r$. To see this, observe that q_j has a factor of N_k whenever $j \neq k$, so that it annihilates all of $U_k = \text{Ker}(N_k)$, thus justifying the second alternative of Eq. 10.4. Using this and Eq. 10.3, we find moreover that

$$u_k = \sum_{j=1}^{r} [h_j(T)q_j(T)](u_k)$$
$$= [h_k(T)q_k(T)](u_k)$$

which justifies the first alternative.

Now suppose that $u_1 + \cdots + u_r = 0$. Then in light of Eq. 10.4, for each index j, applying $h_j(T)q_j(T)$ to both sides shows that $u_j = 0$. Hence all of the summands are zero, as required. Thus point (b) also holds, and V is indeed the direct sum of the characteristic subspaces $U_1,...,U_r$.

(iii) Let T_j denote the restriction of T to an endomorphism of U_j and let μ be an eigenvalue of T_j. Then by Exercise 29 of Chapter 9, $(\mu - \lambda_j)^{m_j}$ is an eigenvalue of the endomorphism $(T_j - \lambda_j 1_{U_j})^{m_j}$. But by construction this latter map is zero on U_j, whence its only eigenvalue is zero. It follows that $\mu = \lambda_j$, as claimed. (The scalar λ_j is itself an eigenvalue of the restricted map since, as we observed earlier, U_j contains the entire eigenspace corresponding to λ_j.)

(iv) Let $n_j = \text{Dim}(U_j)$ and let B_j denote a basis for U_j ($j = 1,...,r$). Then combining the elements of $B_1,...,B_r$ in order, we have a basis B for V with respect to which the matrix of T takes the form

$$A = \begin{pmatrix} A_1 & 0 & \cdots & 0 \\ 0 & A_2 & \cdots & 0 \\ \vdots & \vdots & & \vdots \\ 0 & 0 & \cdots & A_r \end{pmatrix}$$

where each A_j is the $n_j \times n_j$ matrix representing T_j (as defined above) with respect to the basis B_j. The characteristic polynomial of T on V is thus the product of the characteristic polynomials of these restricted maps. By part (iii), T_j has λ_j as its only eigenvalue. Since by assumption all of the roots of the characteristic polynomial of T lie in k, this implies that the characteristic polynomial of T_j is exactly $(t - \lambda_j)^{n_j}$. Thus given our initial factorization (Eq. 10.2) of $p(t)$, the characteristic polynomial of T on V, we have that

$$\prod_{j=1}^{r}(t-\lambda_j)^{m_j} = \prod_{j=1}^{r}(t-\lambda_j)^{n_j}$$

By unique factorization of polynomials, this implies that $m_j = n_j$ ($j=1,\ldots,r$), and this completes the proof.   ❑

10.4   Nilpotent Mappings and the Jordan Normal Form

The aim of this final section is to refine our results on reduction to triangular form. We first extract an easy consequence of Theorem 10-9. This leads us to the examination of nilpotent mappings (defined below) and then to the Jordan normal form (Theorem 10-18).

In preparation we introduce an elementary notion. We call a triangular matrix *supertriangular* if its diagonal entries are 0. Thus a supertriangular matrix takes the form

$$A = \begin{pmatrix} 0 & a_{12} & a_{13} & \cdots & a_{1n} \\ 0 & 0 & a_{23} & \cdots & a_{2n} \\ \vdots & \vdots & \vdots & & \vdots \\ 0 & 0 & 0 & \cdots & a_{n-1,n} \\ 0 & 0 & 0 & \cdots & 0 \end{pmatrix}$$

Supertriangular matrices have the following fundamental property:

10-10 PROPOSITION. *Let* $A \in M_n(k)$ *be supertriangular. Then* $A^n = 0$.

PROOF. At least three strategies come to mind. We could make an argument by induction, based on direct calculation. We could interpret A as the matrix of a linear transformation T on k^n and then argue that as j increases, T^j has monotonically decreasing rank and must eventually become the zero map. But to shamelessly indulge—pardon the split infinitive and the oxymoron to follow—an elegant overkill, nothing surpasses this: The matrix $tI_n - A$ is clearly triangular, with every diagonal entry equal to the indeterminate t. Hence the characteristic polynomial of A is precisely t^n, and therefore, by the Cayley-Hamilton Theorem, one concludes that $A^n = 0$.   ❑

A square matrix A is called *nilpotent* if $A^m = 0$ for some positive integer m. The least such m is called the *index of nilpotency*. (The terminology also applies, in the obvious way, to vector space endomorphisms.) Accordingly, the previous proposition may be paraphrased as follows: a supertriangular $n \times n$ matrix is nilpotent with index of nilpotency bounded by n. (So, too, for an endomorphism represented by such a matrix.)

Characteristic Subspaces Again: Scalar + Nilpotent

We now take a more careful look at the restriction of an endomorphism to a characteristic subspace. Assume that V is a vector space over k of dimension n. Let T be an endomorphism of V such that all of the roots of the characteristic polynomial of T lie in k. As previously, we denote these roots $\lambda_1,\ldots,\lambda_r$ with corresponding multiplicities $m_1,\ldots,m_r$ and corresponding characteristic subspaces $U_1,\ldots,U_r$. Once again we let $T_j: U_j \rightarrow U_j$ denote the restriction of T to an endomorphism of the characteristic subspace U_j $(j=1,\ldots,r)$.

According to Theorem 10-4, Each U_j admits a basis B_j with respect to which the matrix A_j of T_j is triangular. Now the eigenvalues of a triangular matrix are precisely the diagonal entries, but by Theorem 10-9, part (iii), the only eigenvalue for T_j is λ_j. It follows that A_j has the form

$$A_j = \begin{pmatrix} \lambda_j & * & \cdots & * & * \\ 0 & \lambda_j & \cdots & * & * \\ \vdots & \vdots & & \vdots & \vdots \\ 0 & 0 & \cdots & \lambda_j & * \\ 0 & 0 & \cdots & 0 & \lambda_j \end{pmatrix}$$

Clearly A_j may be written as the sum of the scalar matrix $\lambda_j I_{m_j}$ and the supertriangular (hence nilpotent) matrix represented by the asterisks. We have thus proven the following result:

10-11 THEOREM. *The restriction of T to the characteristic subspace U_j may be decomposed as*

$$T_j = D_j + N_j$$

where D_j is the scalar map $\lambda_j I_{m_j}$ and N_j is nilpotent.   □

D_j is called the *diagonal part of T_j*, while N_j is called the *nilpotent part*. Note that since D_j is uniquely defined by T_j (by the eigenvalue λ_j), so is N_j (as the difference between T_j and D_j).

The Structure of Nilpotent Mappings

Theorem 10-11 calls our attention to nilpotent endomorphisms. Let V be a vector space over k of dimension n, and let N be a nilpotent endomorphism of V, so that eventually N^m is the zero map. We shall now construct a special basis for V with respect to which the matrix of N is perfectly simple. To this end, the following result is most fundamental.

10-12 PROPOSITION. *Let $w_0 \in V$, $w_0 \neq 0$, and let m be the least positive integer such that $N^m(w_0) = 0$. Then the vectors*

$$N^{m-1}(w_0), N^{m-2}(w_0), \ldots, N(w_0), w_0$$

are linearly independent.

PROOF. Assume that

$$\sum_{j=1}^{m} \mu_j N^{m-j}(w_0) = 0$$

Then applying N^{m-1} to both sides annihilates all but the last term, showing that μ_m is zero. Next applying N^{m-2} to both sides shows similarly that μ_{m-1} is zero. Continuing in this way, we see that all of the coefficients are zero.   ☐

In the context of this proposition, the subspace W spanned by the linearly independent collection of vectors $N^{m-1}(w_0), N^{m-2}(w_0), \ldots, N(w_0), w_0$ is called a *cyclic subspace* of V with respect to N. We call the indicated basis a *cyclic basis* and call w_0 the *root* of the cyclic basis. Evidently such a subspace is N-invariant and the matrix of N with respect to $N^{m-1}(w_0), N^{m-2}(w_0), \ldots, N(w_0), w_0$ is the $m \times m$ matrix

$$\begin{pmatrix} 0 & 1 & 0 & \cdots & 0 \\ 0 & 0 & 1 & \cdots & 0 \\ 0 & 0 & 0 & \cdots & 0 \\ \vdots & \vdots & \vdots & & 1 \\ 0 & 0 & 0 & \cdots & 0 \end{pmatrix}$$

While the full space V may itself not be cyclic [i.e., there may be no vector v_0 such that the vectors $N^j(v_0)$ ($j = 0, \ldots, n-1$) constitute a basis for V—see Exercise 22 below], we shall show that nonetheless there does exist a basis B for V such that the matrix of N with respect to B has this same beautiful form. The key is to show that V is at least the direct sum of cyclic subspaces. This requires some preliminary work, which we condense into three gentle lemmas.

10-13 LEMMA. *Let W be an m-dimensional cyclic subspace of V with cyclic basis $N^{m-1}(w_0), N^{m-2}(w_0), \ldots, N(w_0), w_0$. Then every element of W may be expressed in the form $q(N)(w_0)$ for some polynomial $q(t)$ in $k[t]$.*

PROOF. Since a basis is a spanning set, every element w in W takes the form

$$w = \sum_{j=1}^{m} \mu_j N^{m-j}(w_0)$$

Thus $w = q(N)(w_0)$ where

$$q(t) = \sum_{j=1}^{m} \mu_j t^{m-j}$$

This completes the proof. □

10-14 LEMMA. *Let W be as above and suppose that for some $q(t) \in k[t]$, $q(N)$ is the zero map on W. Then t^m is a factor of $q(t)$.*

PROOF. We can certainly express $q(t)$ as

$$q(t) = q_1(t)t^m + r(t)$$

for some $q_1(t), r(t) \in k[t]$, where $r(t)$ has degree less than m. (The polynomial $r(t)$ consists of nothing more than the terms of $q(t)$ of degree less than m.) Since N^m is by construction the zero map on W, it follows, in particular, that

$$q(N)(w_0) = [q_1(N)N^m](w_0) + r(N)(w_0)$$
$$= r(N)(w_0)$$

where w_0 is the root of the cyclic basis given for W. Thus if $q(N)$ is also the zero map on W, then $r(N)(w_0) = 0$. But unless r is zero polynomial, this contradicts the linear independence of the basis vectors $N^{m-1}(w_0), N^{m-2}(w_0), \ldots, N(w_0), w_0$. Hence $q(t) = q_1(t)t^m$, and t^m is a factor of $q(t)$, as required. □

10-15 LEMMA. *Let N be a nilpotent endomorphism of V and suppose that V admits a subspace W satisfying the following two conditions:*

(i) *$N(W) = N(V)$; that is, the image of W under N is identical to the image of V under N.*

(ii) *W is the direct sum of subspaces that are cyclic with respect to N.*

Then V is likewise the direct sum of cyclic subspaces.

PROOF. Let $v \in V$. By condition (i) there exists $w \in W$ such that $N(v) = N(w)$. Accordingly, $v - w \in \mathrm{Ker}(N)$, showing that $V = W + \mathrm{Ker}(N)$. Let B_W and B_K denote bases for W and $\mathrm{Ker}(N)$, respectively. By Proposition 4-7, there exists a subset B_K' of B_K such that $B_W \cup B_K'$ is a basis for V. Let W' denote the span of B_K', so that $V = W \oplus W'$. Now consider each of these two summands: The first, W, is the direct sum of cyclic subspaces by assumption (ii). The second, W', is either trivial or, we claim, again the direct sum of cyclic subspaces. Why? Because every nonzero element of Ker(N)—and in particular every element of B'_k—generates a one-dimensional cyclic subspace with itself as cyclic basis! Thus $V = W \oplus W'$ is also the direct sum of cyclic subspaces, as required.   □

This brings us to our main result. The proof is somewhat subtle and bears close attention.

10-16 THEOREM. *Let V be a nonzero vector space and let N be a nilpotent endomorphism of V. Then there exist cyclic subspaces $V_1, \ldots, V_s$ of V with respect to N such that*

$$V = V_1 \oplus V_2 \oplus \cdots \oplus V_s$$

That is, V is the direct sum of cyclic subspaces.

PROOF. The proof goes by induction on n, the dimension of V. The case $n = 1$ is trivial, since a nilpotent map on a one-dimensional vector space must be the zero map (exercise), with respect to which any nonzero element becomes a cyclic basis. So we assume that $n > 1$. The strategy is to construct a subspace which satisfies the hypothesis of the preceding lemma.

First note that $N(V) \neq V$, since otherwise $N^m(V) = V$ for all m, contradicting the assumption that N is nilpotent. Thus by induction, $N(V)$ is the direct sum of cyclic subspaces $W_1, \ldots, W_r$. Let $w_j \in W_j$ ($j = 1, \ldots, r$) denote the root of a cyclic basis for W_j, and let $v_1, \ldots, v_r$ denote, respectively, a collection of pre-images for $w_1, \ldots, w_r$ under N [i.e., $N(v_j) = w_j$ for all j]. Then define W'_j ($j = 1, \ldots, r$) to be the span of W_j together with v_j and define $W' = W'_1 + \cdots + W'_r$. Now clearly each W'_j is cyclic. (Indeed, v_j is the root of a cyclic basis.) Also $N(W') = N(V)$, since by construction the mapping N sends the cyclic basis for W'_j rooted at v_j to the cyclic basis for W_j rooted at w_j ($j = 1, \ldots, r$). Hence in order to conclude the argument by application of Lemma 10-15, it suffices to show that $W' = W'_1 \oplus \cdots \oplus W'_r$, and for this we need only establish the following conditional:

$$\sum_{j=1}^{r} w'_j = 0 \ (w'_j \in W'_j, j = 1, \ldots, r) \Rightarrow w'_j = 0 \ (j = 1, \ldots, r) \qquad (10.5)$$

By Lemma 10-13, $w'_j = q_j(N)(v_j)$ ($j = 1, \ldots, r$) for some family of polynomials $q_1(t), \ldots, q_r(t)$ in $k[t]$, so that the indicated summation takes the form

$$\sum_{j=1}^{r} q_j(N)(v_j) = 0$$

Applying N to both sides and noting that $N(v_j) = w_j$ yields

$$\sum_{j=1}^{r} q_j(N)(w_j) = 0$$

Now $N(V)$ *is the direct sum of cyclic subspaces* $W_1, \ldots, W_r$, whence $q_j(N)(w_j) = 0$ for all j. Since w_j is the root of a cyclic basis for W_j, it follows further that $q_j(N)$ is the zero map on W_j. By a weak application of Lemma 10-14, this shows that t is a factor of each of the $q_j(t)$, which is to say that $q_j(t) = q_j'(t)t$. Therefore

$$\sum_{j=1}^{r} q_j(N)(v_j) = \sum_{j=1}^{r} [q_j'(N)N](v_j)$$

$$= \sum_{j=1}^{r} q_j'(N)(w_j)$$

$$= 0$$

Again applying Lemma 10-14 (this time at full strength), we deduce that for each index j, the monomial t^{m_j} is a factor of $q_j'(t)$, where m_j is the dimension of W_j. Therefore t^{m_j+1} is a factor of $q_j(t)$. But then $q_j(N)$ contains a factor of N^{m_j+1}, which by construction is zero on V_j. Consequently $w_j' = q_j(N)(v_j) = 0$ for all j, as required by the conditional statement 10.5. This completes the proof.   ☐

Theorem 10-16 immediately allows us to extend the simple representation of a nilpotent endomorphism on cyclic subspaces to the full space:

10-17 THEOREM. *Let N be a nilpotent endomorphism of V. Then there exists a basis for V with respect to which the matrix of N has the form*

$$\begin{pmatrix} 0 & 1 & & & & & & \\ & 0 & \ddots & & & & 0 & \\ & & \ddots & 1 & & & & \\ & & & 0 & & & & \\ \hline & & & & \ddots & & & \\ & & & & & 0 & 1 & \\ & 0 & & & & & 0 & \ddots \\ & & & & & & & \ddots & 1 \\ & & & & & & & & 0 \end{pmatrix}$$

PROOF. By the preceding theorem, we can decompose V into the direct sum of cyclic (hence N-invariant) subspaces. By definition, each of these summands admits a basis with respect to which the restricted endomorphism has the given form. Accordingly, the full matrix with respect to the combined basis has such blocks strung along the diagonal, yielding precisely the desired structure. (Note that these square subblocks are not necessarily of the same size; also this form includes the possibility of the zero matrix, when all blocks are of size 1×1.)   ☐

And now the *finale* (marked *allegro con brio*). We combine Theorems 10-11 and 10-17 to obtain an elegant representation of a general endomorphism (still under the assumption that the roots of its characteristic polynomial lie in the field k).

10-18 THEOREM. (Reduction to Jordan Normal Form) *Let V be a vector space over k of dimension n, and let T be an endomorphism of V such that all of the roots of the characteristic polynomial of T lie in k. Assume that these (distinct) roots are $\lambda_1,...,\lambda_r$ with respective multiplicities $m_1,...,m_r$. Then there exists a basis for V with respect to which the matrix of T has block form*

$$A = \begin{pmatrix} A_1 & 0 & \cdots & 0 \\ 0 & A_2 & \cdots & 0 \\ \vdots & \vdots & & \vdots \\ 0 & 0 & \cdots & A_r \end{pmatrix}$$

where A_j is the $m_j \times m_j$ matrix given by

Note that this includes the possibility of a diagonal matrix, again when all blocks are of size 1×1. The subblocks appearing in A_j are called the *basic Jordan blocks* belonging to the eigenvalue λ_j. If these basic blocks occur in order of decreasing size, we then speak of the *Jordan normal form* of A, which is thus well defined up to the ordering of the A_j themselves.

PROOF. We decompose V into its characteristic subspaces $U_1,...,U_r$ with respect to T. By Theorem 10-11, on each of these subspaces T decomposes into the sum of its diagonal part (scalar multiplication by the corresponding eigenvalue) and its nilpotent part. By Theorem 10-17, there is a basis for each characteristic subspace U_j with respect to which the nilpotent part is represented by a string of 1's above the diagonal, with 0's elsewhere. Since the diagonal part on U_j is represented by the scalar matrix $\lambda_j I_{m_j}$ relative to *any* basis, we obtain the blocks as shown.   □

Exercises

1. Let V be a vector space over k. Show that $\mathrm{Hom}(V,V)$ is a commutative ring (with respect to addition and composition of functions) if and only if V has dimension 1.

2. Verify the Cayley-Hamilton Theorem for the matrix

$$A = \begin{pmatrix} 1 & 5 \\ 4 & 2 \end{pmatrix}$$

by direct calculation.

3. Repeat the previous problem for the general 2×2 matrix

$$A = \begin{pmatrix} a & b \\ c & d \end{pmatrix}$$

4. Show that $\mathbf{R}^2$ has no invariant subspaces under left multiplication by the matrix

$$A = \begin{pmatrix} 2 & -1 \\ 5 & -2 \end{pmatrix}$$

5. Show that the invertible upper triangular matrices constitute a subgroup of $GL_n(k)$. (*Hint*: Proceed by induction on n.)

6. Derive a straightforward method for inverting a triangular matrix. Show in particular that the diagonal entries of the inverse matrix are exactly the inverses of the corresponding diagonal entries of the original matrix.

7. Reduce the following matrix in $M_2(\mathbf{R})$ to triangular form:

$$A = \begin{pmatrix} 3 & 2 \\ -2 & -1 \end{pmatrix}$$

8. Let a be a positive real number. Show that the polynomial $t^2 + a$ is irreducible in $\mathbf{R}[t]$ but not in $\mathbf{C}[t]$.

9. The polynomials

$$g_1(x) = x^2 - 5x + 6 \quad \text{and} \quad g_2(x) = x^3 - 1$$

in $\mathbf{R}[x]$ have no nonconstant common factors. Find polynomials $h_1(x)$ and $h_2(x)$ in $\mathbf{R}[x]$ such that

$$g_1(x)h_1(x) + g_2(x)h_2(x) = 1$$

(Such polynomials must exist according to the preliminary remarks in Section 10.3.)

10. Consider the endomorphism T_A defined on $\mathbf{R}^3$ by the matrix

$$A = \begin{pmatrix} 1 & 2 & 1 \\ 0 & 1 & 1 \\ 0 & 0 & 2 \end{pmatrix}$$

Find the characteristic subspaces of T_A. Verify by direct calculation that they are indeed invariant subspaces and effect a direct sum decomposition of $\mathbf{R}^3$.

11. Give an example of a singular matrix which is not nilpotent. (*Hint:* One well-placed nonzero entry will do it.)

12. Let A be a lower triangular matrix with zeros on the diagonal. Show that A is nilpotent. (*Hint:* Show first that a matrix is nilpotent if and only if its transpose is.)

13. Show that if A is a nilpotent matrix and B is similar to A, then B is likewise nilpotent.

14. Show that for any n, the only matrix in $M_n(k)$ that is both diagonal and nilpotent is the zero matrix.

15. Prove that the characteristic polynomial of a nilpotent matrix $A \in M_n(k)$ must be t^n. [*Hint:* Use the following fact from algebra: Given a nonconstant polynomial $p(t)$ in $k[t]$, there exists a field K containing k such that $p(t)$ factors into linear polynomials in $K[t]$. Apply this to the characteristic polynomial of A, and then regard A as an element of $M_n(K)$. A is still nilpotent. What can its eigenvalues be? What must the linear factorization of its characteristic polynomial be?]

16. Let $A \in M_n(k)$ be a nilpotent matrix. Show that the index of nilpotency is less than or equal to n. (*Hint:* The previous problem might apply. Also the proof of Proposition 10-10 suggests a more elementary approach.)

17. Let $v_1,\ldots,v_n$ be a basis for the vector space V and let T be an endomorphism of V. Suppose that there exist nonnegative integers $r_1,\ldots,r_n$ such that

$$T^{r_j}(v_j) = 0 \quad (j = 1,\ldots,n)$$

Show that T is nilpotent.

18. Give an example of a vector space V and an endomorphism T of V such that T is *not* nilpotent but nevertheless satisfies the following condition:

for each $v \in V$ there exists an $n \geq 0$ such that $T^n(v) = 0$

In other words, find an endomorphism T of some space V such that any given vector in V is annihilated by some power of T, but no single power of T annihilates every vector simultaneously. (*Hint:* The previous problem implies that V must be infinite dimensional. Keep this—and your first course in calculus—in mind.)

19. Find the index of nilpotency of the following nilpotent matrix:

$$A = \begin{pmatrix} -1 & 0 & 1 \\ 4 & 2 & -4 \\ 1 & 1 & -1 \end{pmatrix}$$

20. Let $T = T_A$ be the endomorphism of $\mathbf{R}^3$ defined by the matrix A given in the previous problem. Find a cyclic basis for $\mathbf{R}^3$ with respect to T and then verify the change of basis formula as it applies to this new basis and the canonical basis. [In other words, show that when you conjugate the rather tame-looking matrix

$$\begin{pmatrix} 0 & 1 & 0 \\ 0 & 0 & 1 \\ 0 & 0 & 0 \end{pmatrix}$$

by the appropriate transition matrix (from the canonical basis to your cyclic basis), the original matrix A indeed results.]

21. Repeat the previous two exercises for the nilpotent matrix

$$A = \begin{pmatrix} 1 & 1 & 1 \\ -5 & -4 & -4 \\ 4 & 3 & 3 \end{pmatrix}$$

22. Let V be a finite-dimensional vector space, and let N be a nilpotent endo-morphism of V. Show that V is cyclic with respect to N (i.e., admits a cyclic basis) if and only if N has index of nilpotency equal to the dimension of V. [*Hint*: If N has index of nilpotency m, then $N^{m-1}(v) \neq 0$ for some $v \in V$. Hence if $m = \dim(V)$, Proposition 10-12 may be brought to bear. Now what if $m < \dim(V)$? Can a cyclic basis for V exist?]

23. Let N be the nilpotent endomorphism of $\mathbf{R}^3$ defined by the following matrix (with respect to the canonical basis):

$$\begin{pmatrix} 0 & 0 & 1 \\ 0 & 0 & 0 \\ 0 & 0 & 0 \end{pmatrix}$$

Decompose $\mathbf{R}^3$ into the sum of cyclic subspaces with respect to N. (This problem, although computationally trivial, does encompass the key ideas of the proof of Theorem 10-16.)

24. Let N be a nilpotent transformation on the vector space V, and let W be a one-dimensional subspace of V. Show that W is cyclic with respect to N if and only if W lies in $\mathrm{Ker}(N)$.

25. Let V be a vector space of dimension 4, and let N be a nilpotent endo-morphism of V with index of nilpotency equal to 2. Show that V is the direct sum of either 2 or 3 cyclic subspaces. Give specific examples to show that both possibilities in fact do occur.

26. Let $A \in M_7(\mathbf{R})$ have the characteristic polynomial

$$p(t) = (t-1)(t-2)^4(t-5)^2$$

Describe *all* possible Jordan normal forms that might arise from such an endomorphism.

27. Let $A, B \in M_n(k)$ be similar matrices and assume that the roots of their common characteristic polynomial lie in k. Show that A and B have the same Jordan normal form up to the ordering of the Jordan blocks.

Supplementary Topics

The first section of this supplementary chapter ties linear algebra into calculus and explains some fundamental aspects of multivariable differentiation. In the second section, we revisit the determinant of real matrices in order to introduce a surprising geometric interpretation. This yields a fresh explanation of the multiplicative nature of the determinant map and accounts for at least the plausibility of the change of variables formula in multivariable calculus. The third section briefly introduces quadratic forms, an idea with many applications but not otherwise required in our main exposition. The final section is sheer extravagance: an introduction to categories and functors, two modern notions that have fundamentally changed the language and nature of mathematical inquiry. A functional understanding of this topic requires a larger base of mathematical experience than most students will have at this point, but perhaps it is not too soon at least to broach the topic.

The results here are for the most part offered without proof and without exercises. The aim is not to develop technical facility but to broaden one's mathematical perspective and mathematical culture.

1 Differentiation

Let f be a real-valued function defined on an open interval $I \subseteq \mathbf{R}$ and let a be an element of I. Recall that the derivative $f'(a)$ is defined by the formula

$$f'(a) = \lim_{h \to 0} \frac{f(a+h) - f(a)}{h} \tag{1}$$

provided that the indicated limit exists. If so, we say that f is *differentiable at a*. In this section we shall explain how differentiation generalizes to a multivariable function $f: U \to \mathbf{R}^m$ defined on an open subset U of $\mathbf{R}^n$. To do so directly via Eq. 1 would require that we replace the real numbers a and h by vectors, while recognizing that the values of f must now also be read as vectors. But this leads immediately to nonsense: the quotient of an m-dimensional vector by an n-dimensional vector. Clearly we must find an alternate path.

Since the obstruction to the generalization of Eq. 1 is division, a sensible approach to the problem is to reformulate this equation without the quotient. This is not difficult if we remember that the derivative $f'(a)$ also provides the best possible linear approximation to f near a.

PROPOSITION. *Suppose that f is differentiable at a. Then there exists a function g defined on an open interval containing 0 such that the following two conditions hold:*

(i) $f(a+h) = f(a) + f'(a) \cdot h + |h|g(h)$

(ii) $\lim_{h \to 0} g(h) = 0$

PROOF. For $h \neq 0$, we define g by solving the first equation for $g(h)$. Thus

$$g(h) = \frac{f(a+h) - f(a) - f'(a) \cdot h}{|h|} \qquad (h \neq 0)$$

Now define $g(0) = 0$. This construction guarantees the existence of a function g that at least satisfies (i). Part (ii) is almost equally trivial. Multiplying our last equation by $|h|$, we have

$$|h| \cdot g(h) = f(a+h) - f(a) - f'(a) \cdot h$$

Dividing both sides by $h \neq 0$ yields

$$\text{sgn}(h) \cdot g(h) = \frac{f(a+h) - f(a)}{h} - f'(a)$$

where sgn denotes the sign function (+1 for positive arguments, −1 for negative arguments). The limit of the right-hand side as h approaches zero is itself zero by definition of the derivative. Hence $g(h)$ must likewise have limit zero as h approaches zero, precisely as claimed.   □

Define a linear transformation $\lambda : \mathbf{R} \to \mathbf{R}$ by $\lambda(h) = f'(a) \cdot h$. Then the equation given in part (i) of the proposition may be recast as follows:

$$f(a+h) = f(a) + \lambda(h) + |h|g(h)$$

We see the right-hand side as the sum of a constant, a linear term, and a term that vanishes rapidly as h approaches 0. The key points in this formulation are that there is no division and that the derivative has been identified with a linear map. Now we are ready to generalize.

First recall that a subset U of $\mathbf{R}^n$ is called *open* if for every $a \in U$ there is a positive real number ε (which may depend on a) such that whenever $|x - a| < \varepsilon$ for a vector $x \in \mathbf{R}^n$, then x also lies in U; that is, U is open if all points in $\mathbf{R}^n$ sufficiently close to a given point of U likewise lie in U. (In the present context, distance is measured with respect to the standard inner product on $\mathbf{R}^n$.) We need this notion of an open set so that the limits appearing below make sense.

DEFINITION. Let $f: U \to \mathbf{R}^m$ be a function defined on an open subset U of $\mathbf{R}^n$. Then given a point $a \in U$, we say that f is *differentiable at* a if there exists a linear transformation $\lambda: \mathbf{R}^n \to \mathbf{R}^m$ such that

$$f(a+h) = f(a) + \lambda(h) + |h|g(h) \tag{2}$$

for some function g defined on a neighborhood of $0 \in \mathbf{R}^n$ (also with codomain $\mathbf{R}^m$) satisfying the condition $\lim_{h \to 0} g(h) = 0$.

One can show that if a linear transformation λ satisfying this definition exists, it is also unique. It is therefore appropriately called *the derivative of f at* a and denoted $Df(a)$.

We know from Chapter 6 that wherever it is defined, $Df(a)$ may be realized by an $m \times n$ matrix (relative to the canonical bases for $\mathbf{R}^n$ and $\mathbf{R}^m$). In a moment we shall see how to compute this matrix, but first we recast Eq. 2 in matrix form as follows:

$$f(a+h) = f(a) + Df(a)h + |h|g(h) \tag{3}$$

Note that the middle term on the right now represents the product of an $m \times n$ matrix with a column vector in $\mathbf{R}^n$. Again $\lim_{h \to 0} g(h) = 0$.

To compute $Df(a)$ using the standard techniques of calculus, we introduce a simple device. If f is any function with codomain $\mathbf{R}^m$, we can express the value of f at any point a of its domain in the following form:

$$f(a) = \begin{pmatrix} f_1(a) \\ f_2(a) \\ \vdots \\ f_m(a) \end{pmatrix}$$

The real-valued functions $f_j = \rho_j \circ f$ $(j = 1, \ldots m)$ are called the *component functions* of f, and each has the same domain as f.

The following theorem is the fundamental result on differentiation of multivariable functions. We state it without proof.

THEOREM. *Let $f: U \to \mathbf{R}^m$ be a function defined on an open subset U of $\mathbf{R}^n$.*

(i) *If f is differentiable at some point $a \in U$, each of the partial derivatives*

$$\frac{\partial f_i}{\partial x_j} \quad (1 \le i \le m, 1 \le j \le n)$$

exists at a; moreover, the derivative of f at a is given by the $m \times n$ matrix

$$Df(a) = \left(\frac{\partial f_i}{\partial x_j}(a) \right)$$

(ii) *Suppose conversely that each of the indicated partials exists in an open neighborhood of a and is continuous at a. Then f is differentiable at a.*

The matrix appearing in Eq. 3 is called the *Jacobian matrix of f*. For functions built from the familiar elementary functions of calculus, it is trivial (if sometimes tedious) to compute.

SPECIAL CASES

(1) Consider a function $f: \mathbf{R} \to \mathbf{R}^m$. Then f is called a *path* in $\mathbf{R}^m$, since if we think of the independent variable as time, f describes a trajectory in m-space. (Hence in this case we prefer to use the variable t rather than x.) Expressing f in terms of its component functions

$$f(t) = \begin{pmatrix} f_1(t) \\ f_2(t) \\ \vdots \\ f_m(t) \end{pmatrix}$$

we compute the *path derivative* by direct differentiation of each of the components:

$$Df(t) = \begin{pmatrix} f_1'(t) \\ f_2'(t) \\ \vdots \\ f_m'(t) \end{pmatrix}$$

If f describes position, Df describes velocity.

(2) Consider a function $f: \mathbf{R}^n \to \mathbf{R}$. Then the derivative of f takes the form of a row vector that is called the *gradient* of f and denoted grad f:

$$Df(a) = (\operatorname{grad} f)(a)$$

$$= \left(\frac{\partial f}{\partial x_1}(a), \frac{\partial f}{\partial x_2}(a), \ldots, \frac{\partial f}{\partial x_n}(a) \right)$$

In this case, Eq. 3 yields the linear approximation

$$f(a+h) \approx f(a) + \operatorname{grad} f(a) \cdot h$$

$$\approx f(a) + \sum_{j=1}^{n} \frac{\partial f}{\partial x_j}(a) \cdot h_j$$

for small h. If we rewrite this in the slightly different form

$$f(a+h) - f(a) \approx \sum_{j=1}^{n} \frac{\partial f}{\partial x_j}(a) \cdot h_j$$

it explains why df, the so-called *total differential* of f, is often expressed as follows:

$$df = \frac{\partial f}{\partial x_1} dx_1 + \frac{\partial f}{\partial x_2} dx_2 + \cdots + \frac{\partial f}{\partial x_n} dx_n$$

(3) As our final special case, we consider a matrix $A \in \mathrm{Mat}_{m \times n}(\mathbf{R})$ and the associated linear transformation $T_A: \mathbf{R}^n \to \mathbf{R}^m$ defined as left multiplication by A. We invite the reader to show that T_A is differentiable with derivative equal to A itself. This generalizes a familiar triviality from one-variable calculus, namely that $d(ax)/dx = a$.

   To summarize the main point of this section, the derivative of a function at a given point is a number or a matrix only insofar as this number or matrix represents a linear transformation that may be used to approximate the function locally. In this sense, the notion of derivative that one usually learns in a first course in calculus is slightly misleading, although entirely forgivably so.

2 The Determinant Revisited

We shall now introduce a surprising interpretation of the determinant of a real $n \times n$ matrix as a measure of volume. Since the determinant of a real matrix may

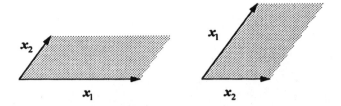

Figure 1. On the left, x_1 may be rotated onto x_2 by a counterclockwise rotation of less than π radians. On the right, this is clearly not possible.

be negative, we shall first have to introduce the notion of *oriented volume* or *oriented area*. The latter term is appropriate in two-dimensional settings, where we shall illustrate these ideas.

Consider two linearly independent vectors in x_1 and x_2 in $\mathbf{R}^2$, which therefore constitute an ordered basis. (In this context the ordering is paramount!) Figure 1 shows how these vectors may relate to each other. If it is possible to rotate x_1 counterclockwise by less than π radians onto x_2, we shall say that the pair of vectors has positive orientation. Otherwise, we shall call the associated orientation negative. In both cases the vectors span a parallelogram, as shown. Let us define a real-valued function $\alpha(x_1,x_2)$ on $\mathbf{R}^2$ as follows:

If x_1 and x_2 are linearly independent and have positive orientation, then $\alpha(x_1,x_2)$ is just the area of the parallelogram spanned by these vectors.

If x_1 and x_2 are linearly independent and have negative orientation, then $\alpha(x_1,x_2)$ is minus the area of the parallelogram spanned by these vectors.

If x_1 and x_2 are linearly dependent, then $\alpha(x_1,x_2)$ is zero.

We shall call $\alpha(x_1,x_2)$ the *oriented area* of the parallelogram spanned by the vectors x_1 and x_2.

Let us now deduce some properties of oriented area. Consider the calculation of $\alpha(cx_1,x_2)$ where c is any real number. If c is positive, the parallelogram spanned by x_1 and x_2 has been either stretched or contracted in one direction by a factor of c, with no change in orientation. If c is negative, the parallelogram is again stretched or contracted, but also suffers a reversal of orientation. If c is zero, the parallelogram of course degenerates into a line segment with no area. The upshot of this analysis is that $\alpha(cx_1,x_2) = c\alpha(x_1,x_2)$. The argument works equally well in the second variable.

Figure 2 shows an even more striking property of oriented area: α is also additive in each variable. Combining this with the analysis of the last paragraph, we have

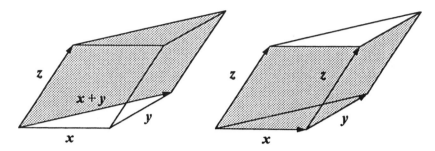

Figure 2. Comparison of the two shaded regions (which differ by translation of a triangle) shows that $\alpha(x+y,z)=\alpha(x,z)+\alpha(y,z)$. A similar analysis holds in the second variable.

(i) *The oriented area function α is linear in each variable.*

And we can immediately state two other properties of oriented area, both of which are rather trivial-looking observations:

(ii) *For all vectors $x \in \mathbf{R}^2$, $\alpha(x, x)=0$.*

(iii) *The oriented area of the parallelogram spanned by the canonical basis vectors e_1 and e_2 is 1; that is, $\alpha(e_1,e_2)=1$.*

Now what can be the importance of these remarks?
  Recall that the Fundamental Theorem of Determinants (8-1) says in part that the determinant is uniquely characterized by three properties identical to those observed here for oriented area. Consequently, we have in splendid ignorance established that for a 2×2 real matrix A, $\det(A)$ is precisely the oriented area of the parallelogram spanned by the columns of A! In particular,

$$|\det(A)| = \text{the area of the parallelogram spanned by the columns of } A$$

This is a result that one would hardly expect from our earlier algebraic characterization of the determinant. (Unless, of course, one did Exercise 28 of Chapter 8.)
  In real n-space, n linearly independent vectors $x_1,\ldots,x_n$ span an n-dimensional object called a *parallelepiped* or *paralleletope*. This is defined as the following subset of $\mathbf{R}^n$:

$$\left\{ \sum_{j=1}^{n} a_j x_j : a_1,\ldots,a_n \in [0,1] \right\}$$

While it is harder to carry our intuitive understanding of orientation into higher dimensions (in fact, the determinant itself is the most convenient tool for such analysis), the results derived for the plane may be extended rigorously. Thus for any $n\times n$ real matrix A, $\det(A)$ is precisely the oriented volume of the parallelepiped spanned by the columns of A. (Or, for that matter, by the rows of A. Do you see why?) In particular,

$|\det(A)|$ = the volume of the parallelepiped spanned by the columns of A

We have thus achieved a complementary, geometric description of the determinant. (Note that the present discussion does not apply over abstract fields; hence our previous algebraic description remains the more fundamental.)

We can immediately put this new geometric interpretation of the determinant to good use. Consider a nonsingular endomorphism T of $\mathbf{R}^n$ as represented by a matrix A with respect to the canonical basis. The columns of A are precisely the images of the canonical basis vectors under the mapping T. Hence the parallelepiped—in this case a hypercube—spanned by $e_1,...,e_n$ maps to the parallelepiped spanned by the columns of A. Since the hypercube has unit volume and its image under T has volume $|\det(A)|$, we can accordingly regard the determinant as a measure of the expansion in volume effected by T. This yields a satisfying geometric explanation for the multiplicativity of the determinant: the application of two successive endomorphisms results in two successive expansions in volume; the cumulative effect is clearly the product of the two.

The determinant as volume makes its appearance in calculus in a famous and fundamental theorem. Before stating it, we recall from the previous section that the derivative of a function g from $\mathbf{R}^n$ to $\mathbf{R}^m$ is a linear transformation Dg also from $\mathbf{R}^n$ to $\mathbf{R}^m$. In particular, the derivative of a function g from $\mathbf{R}^n$ to itself is an endomorphism of $\mathbf{R}^n$, and thus it makes sense to speak of its determinant.

THEOREM. (Change of Variables Formula) *Let U be an open subset of $\mathbf{R}^n$ and suppose that $g: U \rightarrow \mathbf{R}^n$ meets the following conditions:*

(i)   *g is injective;*

(ii) *the derivative Dg exists and is continuous throughout U;*

(iii) *Dg has nonzero determinant throughout U.*

Then if f is a real-valued integrable function on the image of U under g,

$$\int_{g(U)} f(v)dv = \int_U f \circ g(u) \cdot |\det Dg(u)|du$$

We shall, of course, attempt no proof of this theorem, but let us at least point out that the factor $|\det Dg(u)|$ appears on the right side precisely to account for the local expansion that a differential volume element experiences when mapped from U to $g(U)$ via g.

3 Quadratic Forms

Quadratic forms appear in many branches of mathematics, from statistics to number theory. In this section we introduce the subject over an abstract field and then state some of the most useful results about real quadratic forms.

Let V be an n-dimensional vector space over an arbitrary field k. Then a *bilinear form* on V is a mapping $B: V \times V \to k$ that is linear in each variable. More explicitly,

$$B(\lambda u + \mu v, w) = \lambda B(u, w) + \mu B(v, w)$$

for all $\lambda, \mu \in k$ and $u, v, w \in V$, with a similar equation in the second variable. A bilinear form B is *symmetric* if $B(u, v) = B(v, u)$ for all $u, v \in V$.

Let us now fix a basis for V and identify V with k^n via the associated co-ordinate isomorphism. Along the lines of Proposition 7-5, one can easily show that the bilinear map B takes the form

$$B(x, y) = {}^t x A y \quad (x, y \in k^n)$$

for some $A \in M_n(k)$. Moreover, A is symmetric if and only if the bilinear form B is likewise symmetric.

With V as above, a map $Q: V \to k$ is called a *quadratic form* if it satisfies the following conditions:

(i)   for all $u \in V$ and $\lambda \in k$, $Q(\lambda u) = \lambda^2 Q(u)$;

(ii) the map $B_Q: V \times V \to k$ defined by $B_Q(u, v) = Q(u, v) - Q(u) - Q(v)$ is a symmetric bilinear form on V.

Thus every quadratic form Q gives rise to an associated bilinear form B_Q. We can in fact reverse this correspondence subject to one further technical condition (which the reader concerned only with the fields $\mathbf{R}$ and $\mathbf{C}$ may ignore).

A field k is said to be of *characteristic* 2 if $1 + 1 = 0$ in k. (Here 0 and 1 stand, respectively, for the additive and multiplicative identities of the field k, and not necessarily for ordinary integers.) For example, the finite field $\mathbf{F}_2$ has characteristic 2 (cf. Section 2.3), and there are infinitely many other such fields. If a field k is *not* of characteristic 2, we may use the customary symbol 2 to denote the sum $1 + 1$. Then 2 is multiplicatively invertible in k, since by definition k^* is a group. As usual, we denote this inverse $\frac{1}{2}$.

Resuming the general discussion, assume henceforth that k is not of characteristic 2, so that the fraction $\frac{1}{2}$ makes sense in k. Then with an arbitrary symmetric bilinear form B, we may associate the function $Q_B: V \to k$ defined by

$$Q_B(u) = \frac{1}{2} B(u, u)$$

Then Q_B turns out to be a quadratic form. In fact, we have the following bijective correspondence:

PROPOSITION. *Assume that k is not of characteristic 2. Then the mapping*

$$Q \mapsto B_Q$$

constitutes a bijective correspondence between the set of quadratic forms on V and the set of symmetric bilinear forms on V. Moreover, the inverse correspondence is given by

$$B \mapsto Q_B$$

PROOF. We need only show that for any bilinear form B, Q_B is indeed a quadratic form and that the indicated mappings are in fact mutually inverse. First we compute that

$$Q_B(\lambda u) = \tfrac{1}{2} B(\lambda u, \lambda u) = \tfrac{1}{2} \lambda^2 B(u,u) = \lambda^2 Q_B(u)$$

Second we note that

$$\begin{aligned}
Q_B(u+v) - Q_B(u) - Q_B(v) &= \tfrac{1}{2}\big(B(u+v,u+v) - B(u,u) - B(v,v)\big) \\
&= \tfrac{1}{2}\big(B(u,v) + B(v,u)\big) \\
&= B(u,v)
\end{aligned}$$

This shows both that Q_B is indeed a quadratic form and that the first map given in the proposition is at least a left inverse for the second. Finally, we have that

$$\begin{aligned}
\tfrac{1}{2} B_Q(u,u) &= \tfrac{1}{2}\big(Q(u+u) - Q(u) - Q(u)\big) \\
&= \tfrac{1}{2}\big(Q(2u) - 2Q(u)\big) \\
&= \tfrac{1}{2}\big(4Q(u) - 2Q(u)\big) \\
&= Q(u)
\end{aligned}$$

This establishes the inverse relationship in the other direction and thus concludes the proof. ☐

We now pass to a less abstract characterization of quadratic forms on k^n (where k still is not of characteristic 2). Since each such form Q arises in connection with a symmetric bilinear form and every bilinear form is represented by a symmetric matrix with respect to the canonical basis, we can always write Q as

$$Q(x) = {}^t x A x$$
$$= \sum_{i,j} a_{ij} x_i x_j$$

where A is a symmetric $n \times n$ matrix. (Here the indices vary independently from 1 to n, so that there are n^2 terms in the summation.) We shall say that A *represents* Q.

We next consider a change of variables. Suppose that $x = Py$ for some matrix $P \in GL_n(k)$. Let $B = {}^t PAP$, where A represents Q. Define another quadratic form R by $R(y) = {}^t yBy$. Then

$$Q(x) = {}^t x A x$$
$$= {}^t(Py)A(Py)$$
$$= {}^t y({}^t PAP) y$$
$$= {}^t yBy$$
$$= R(y)$$

In other words, except for a linear change in coordinates, A and B represent the same form. This leads to the following definitions:

DEFINITIONS. If $B = {}^t PAP$ for a nonsingular matrix P, then A and B are said to be *congruent*. Two quadratic forms represented by congruent matrices are called *equivalent*.

One shows easily that this is an equivalence relation on $M_n(k)$ which preserves rank. A fundamental issue is then to determine when two forms are equivalent. We shall have a little to say about this in the special case that $k = \mathbf{R}$.

Real Quadratic Forms

A quadratic form on a real vector space is called a *real quadratic form*. Recall that according to the Spectral Decomposition Theorem, a real symmetric matrix is diagonalizable by an orthogonal matrix (for which the operations of transpose and inverse are identical).

THEOREM. (The Principal Axis Theorem) *Let* $Q : \mathbf{R}^n \to \mathbf{R}$ *be a real quadratic form. Then* Q *is equivalent, via an orthogonal transformation, to the form*

$$\lambda_1 y_1^2 + \cdots + \lambda_n y_n^2$$

where $\lambda_1, \ldots, \lambda_n$ *are the eigenvalues of the matrix representing* Q.

PROOF. Suppose that A represents Q. There exists an orthogonal matrix P such that

$$D = \begin{pmatrix} \lambda_1 & 0 & \cdots & 0 \\ 0 & \lambda_2 & \cdots & 0 \\ \vdots & \vdots & \ddots & \vdots \\ 0 & 0 & \cdots & \lambda_n \end{pmatrix} = {}^{t}PAP$$

where $\lambda_1, \ldots, \lambda_n$ are the eigenvalues of A. Clearly the quadratic form represented by the diagonal matrix D is that shown.   ❑

COROLLARY. *More specifically, every real quadratic form is equivalent to one of the type*

$$y_1^2 + \cdots + y_p^2 - y_{p+1}^2 - \cdots - y_{p+q}^2$$

PROOF. We may suppose that the given form has already been diagonalized and is of the type

$$\lambda_1 x_1^2 + \cdots + \lambda_n x_n^2$$

Assuming that $\lambda_1, \ldots, \lambda_p$ are positive while $\lambda_{p+1}, \ldots, \lambda_{p+q}$ are negative, with the other coefficients zero, consider the following further change of variables:

$$x_j = \frac{1}{\sqrt{\lambda_j}} y_j \quad (j = 1, \ldots, p)$$

$$x_j = \frac{1}{\sqrt{-\lambda_j}} y_j \quad (j = p+1, \ldots, p+q)$$

$$x_j = y_j \quad (j = p+q+1, \ldots, n)$$

This clearly transforms $\lambda_1 x_1^2 + \cdots + \lambda_n x_n^2$ into the required expression.   ❑

The ordered pair of integers (p, q) defined in the previous corollary is called the *signature* of the quadratic form. While we shall not prove it, the signature is an invariant of the form, which is to say that equivalent forms have the same signature. In fact, much more is true:

THEOREM. (Sylvester's Law of Inertia) *Two real quadratic forms on real n-space are equivalent if and only if they have the same signature.*

We complete this section with one example of the amazing power of this result.

EXAMPLE

Suppose that a quadratic form Q on $\mathbf{R}^4$ is represented by a matrix A whose eigenvalues are +2, +1, 0, −1. Then Q has signature (2,1) and is therefore equivalent to the form

$$x_1^2 + x_2^2 - x_3^2$$

Moreover, any quadratic form on $\mathbf{R}^4$ whose matrix has two positive eigenvalues and one negative eigenvalue is likewise equivalent to this form and equivalent to Q.

4 An Introduction to Categories and Functors

Samuel Eilenberg and Saunders Mac Lane introduced categories and functors into mathematics in 1945. These concepts represent as great a "triumph of generalization" (to quote John Fowles) as Darwin's theory of evolution. In the latter half of the twentieth century, they have not only changed the language of mathematical discourse, but also the nature of mathematical inquiry.

While category theory, like group theory, has a technical life of its own, we are not concerned here with its technical development. Our only purpose is to exhibit the most basic definitions and examples so that the student may at least glimpse the beauty of these ideas as a framework for mathematics.

DEFINITION (following Lang's *Algebra*, 1965). A *category* $\mathscr{A}$ consists of

(i)   a class of *objects* (not necessarily a set), denoted Ob($\mathscr{A}$);

(ii) for every pair of objects $A,B \in$ Ob($\mathscr{A}$), a set Mor(A,B) of *morphisms* from A to B;

(iii) for every three objects $A,B,C \in$ Ob($\mathscr{A}$), a *law of composition*, which is to say a function Mor(B,C)×Mor(A,B)→Mor(A,C).

Moreover, these data are subject to the following axioms:

CAT 1. Two sets Mor(A,B) and Mor(A',B') are disjoint unless $A=A'$ and $B=B'$.

CAT 2. For each $A \in$ Ob($\mathscr{A}$), there is a morphism $1_A \in$ Mor(A,A) which is neutral with respect to composition.

CAT 3. The law of composition is associative.

To state these last two axioms more precisely, we need some notation. Henceforth we write $f:A\to B$ to indicate that $f\in\mathrm{Mor}(A,B)$ and use $\circ$ to denote the composition operator. Thus if we have $g:B\to C$ and $f:A\to B$, the composed morphism mandated by (iii) above is $g\circ f:A\to C$. Then CAT 2 says that for every object A, there is a morphism $1_A:A\to A$ such that

$$1_A\circ g=g \quad\text{and}\quad h\circ 1_A=h$$

for all morphisms g from any object to A and all morphisms h from A to any object. CAT 3 says that whenever three morphisms f, g, and h admit composition as indicated, we have

$$(h\circ g)\circ f=h\circ(g\circ f)$$

Note that these properties seem modeled on ordinary composition of functions. Indeed, sets and functions provide the first of a few examples likely to be familiar to the reader.

EXAMPLES

(1) *Sets and Functions.* The objects are sets; the morphisms are functions. The composition law is ordinary composition of functions, which we know to be always associative. For any set A, the neutral element in $\mathrm{Mor}(A,A)$ is the identity function on A.

(2) *Topological Spaces and Continuous Mappings.* The objects are topological spaces and the morphisms are continuous mappings between spaces. To see that this is a category, it suffices to note that the identity map is always continuous and that composition of continuous functions is continuous. The other properties then follow from the ordinary rules for composition of functions.

(3) *Groups and Group Homomorphisms.* The objects are groups and the morphisms are group homomorphisms. As for the previous example, it suffices to note that the identity map is a group homomorphism and that the composition of group homomorphisms is a group homomorphism.

(4) *Vector Spaces and Linear Transformations* (over a fixed field). Let k be a field. The class of all vector spaces over k constitutes a category for which the morphisms consist of k-linear transformations. As above, it suffices to note that the identity map is a linear transformation and that composition of linear transformations is a linear transformation.

Now these examples are somewhat restrained insofar as in all cases the objects are sets and the morphisms are functions, although possibly of a restricted

$$A \xrightarrow{\ f\ } B$$

$$\downarrow \qquad \downarrow$$

$$F(A) \xrightarrow{F(f)} F(B)$$

$$A \xrightarrow{\ f\ } B$$

$$\downarrow \qquad \downarrow$$

$$F(A) \xleftarrow{F(f)} F(B)$$

Figure 3. Illustration of both covariant and contravariant functors. The vertical arrows represent assignments of objects; they are not themselves morphisms.

type. Much more abstract categories are possible and useful (e.g., the category of ringed spaces from algebraic geometry). But the point we want to stress is that the morphisms—the structure preserving maps—are part of the defining data. In this sense the categorical approach is strongly relational. This will appear even more strikingly below as we introduce the notion of a functor.

DEFINITION. Let $\mathscr{A}$ and $\mathscr{B}$ be categories. Then a *covariant functor* F from $\mathscr{A}$ to $\mathscr{B}$ is a rule which assigns to every object $A \in \mathrm{Ob}(\mathscr{A})$ an object $F(A) \in \mathrm{Ob}(\mathscr{B})$ and to each morphism $f:A \rightarrow B$ a morphism $F(f):F(A) \rightarrow F(B)$, subject to the following axioms:

FUN 1. For all $A \in \mathrm{Ob}(\mathscr{A})$, $F(1_A)=1_{F(A)}$.

FUN 2. If $f:A \rightarrow B$ and $g:B \rightarrow C$, then $F(g \circ f)=F(g) \circ F(f)$

We also have the notion of a *contravariant functor* F, which assigns objects and morphisms across categories as above, but in this case reverses the direction of a morphism: if $f:A \rightarrow B$, then $F(f):F(B) \rightarrow F(A)$. Accordingly, the second axiom must be amended to read

FUN 2'. If $f:A \rightarrow B$ and $g:B \rightarrow C$, then $F(g \circ f)=F(f) \circ F(g)$

Figure 3 illustrates the relationships of objects and morphisms for both covariant and contravariant functors.

EXAMPLES

(1) Let $\mathscr{A}$ be the category whose objects are finite sets and whose morphisms are ordinary functions, and let $\mathscr{B}$ be the category of vector spaces and linear transformations over some fixed field k. Given any finite set S, we can associate with it a real finite-dimensional vector space $F(S)$ whose basis is S. Thus $F(S)$ consists of formal sums

$$\sum_{s \in S} \lambda_s \cdot s \quad (\lambda_s \in k)$$

where two such expressions are equal if and only if all coefficients match. Addition and scalar multiplication are, as usual, defined componentwise:

$$\sum \lambda_s \cdot s + \sum \mu_s \cdot s = \sum (\lambda_s + \mu_s) \cdot s$$
$$\mu \sum \lambda_s \cdot s = \sum (\mu \lambda_s) \cdot s$$

Any morphism of finite sets (i.e., a function) $f: S \rightarrow T$ induces the following morphism of vector spaces (i.e., a linear transformation) $F(f): F(S) \rightarrow F(T)$ on the associated spaces:

$$\sum \lambda_s \cdot s \mapsto \sum \lambda_s \cdot f(s)$$

This is just extension by linearity, as explained in Section 6.4. Clearly the identity map on a finite set S induces the identity map on the associated vector space $F(S)$. Moreover, the induced functions respect composition:

$$\text{if } f: S \rightarrow T \text{ and } g: T \rightarrow U, \text{ then } F(g \circ f) = F(g) \circ F(f)$$

Hence F constitutes a covariant functor from $\mathscr{A}$ to $\mathscr{B}$. (This may not be a particularly useful functor, but it does illustrate the main technical points.)

(2) Our second example is both interesting and useful and has even appeared previously in the text! Consider the category of vector spaces over a fixed field k. Then to every vector space V we can associate its dual space $V^* = \text{Hom}(V, k)$. Furthermore, if $T: V \rightarrow W$ is a linear transformation, we have seen in Section 6.4 that

$$T^*: W^* \rightarrow V^*$$
$$f \mapsto f \circ T$$

is likewise a linear transformation on the dual spaces. Our comments just prior to Proposition 6-16 say precisely that $\text{Hom}(-, k)$ is a contravariant functor from this category to itself. The reader who has studied the remainder of that section will note the good use made of this functoriality in the subsequent analysis.

Generality and Functoriality

The introduction of categories and functors has, over the last half century, changed the nature of mathematics both explicitly and implicitly. For example, both algebraic topology and algebraic geometry are much driven by the search for functors from topological spaces to discrete algebraic categories (such as groups and rings) that yield decisive information on the classification of spaces.

A less obvious effect occurs at the very heart of mathematics in connection with the process of abstraction: nowadays a definition might be considered suspect if the collection of objects it encompasses does not constitute a category; an association of objects across categories might be held defective if it is not functorial. We close our discussion with a few heuristic comments that may in some small way explain the immense effectiveness and appeal of the categorical framework. We begin with some preliminary remarks on the nature of generality in mathematics.

A natural approach to achieving mathematical generality is through generalized objects or forms. For example, we saw in Section 2.3 that both the integers and the set of continuous real-valued functions defined on the real numbers admit addition and multiplication subject to some very familiar laws. Consequently both are subsumed under the mathematical structure of a commutative ring, which is accordingly an appropriate generalization of each. We speak of and study rings, and whatever their abstract properties, they are shared by the integers and by the continuous real-valued functions. The power of this approach is undeniable, and yet, as we shall now argue, necessarily limited.

Axiomatic systems such as rings, groups, fields, and topological spaces distill gradually out of mathematical experience. They are a means both to unify mathematics and to isolate the key properties of certain well-studied objects. Indeed, the capacity to identify viable forms is a quintessential and rare mathematical talent. To understand why this is more art than science, consider, for example, the notoriously austere axioms for an abstract group or a topological space. These resemble cosmetically any number of simple axiomatic systems that one might define, but their particular richness derives from two attributes, which by nature lie in opposition to each other:

(i)   They are sufficiently general to encompass a wide spectrum of mathematical phenomena.

(ii) They are sufficiently restrictive to capture essential features of some part of the mathematical landscape.

Point (i) alone is clearly insufficient. A magma (a set together with an operation, subject to no restrictions whatsoever) is certainly a more general object than a group, but is it correspondingly more central to mathematics? Of course not; so general an object commands virtually no interest. One might similarly relax the axioms for a topological space to achieve greater inclusiveness—but in this case at the expense of key features of spatiality. Thus the warning signs are clearly posted: transgress point (ii) at your peril.

By stressing relationships *across* classes, functoriality deftly sidesteps the contention described above between generality and richness. Thus it achieves a measure of unity without retreat into banality. Moreover, functorial relationships often allow one to bring to bear the full knowledge of one class of objects on the analysis of another. In this way categories and functors provide within mathematics itself a service that mathematics has long provided for science. And finally, in purely aesthetic terms, the exquisite balance between the objectified and the relational manifest in this approach reveals one of the most beautiful designs ever conceived by the human mind.

Index

Printed in the United States
39049LVS00002B/184-225

9 780387 940991